Quantum Computer Science

Synthesis Lectures on Quantum Computing

Synthesis Lectures On Quantum Computing presents works on research and development in quantum information and computation for an audience of professional developers, researchers and advanced students. Topics include Quantum Algorithms, Quantum Information Theory, Error Correction Protocols, Fault Tolerant Quantum Computing, the Physical Realizations of Quantum Computers, Quantum Cryptography, and others of interest.

Quantum Computer Science
Marco Lanzagorta and Jeffrey Uhlmann
2009

Quantum Walks for Computer Scientists
Salvador Elias Venegas-Andraca
2008

Quantum Computer Science

Marco Lanzagorta and Jeffrey Uhlmann

www.morganclaypool.com

ISBN: 9781598297324 paperback
ISBN: 9781598297331 ebook

DOI 10.2200/S00159ED1V01Y200810QMC002

A Publication in the Morgan & Claypool Publishers series
SYNTHESIS LECTURES ON QUANTUM COMPUTING

Lecture #2
Series Editors: Marco Lanzagorta, ITT Corporation
Jeffrey Uhlmann, University of Missouri-Columbia

Quantum Computer Science

Marco Lanzagorta
ITT Corporation

Jeffrey Uhlmann
University of Missouri-Columbia

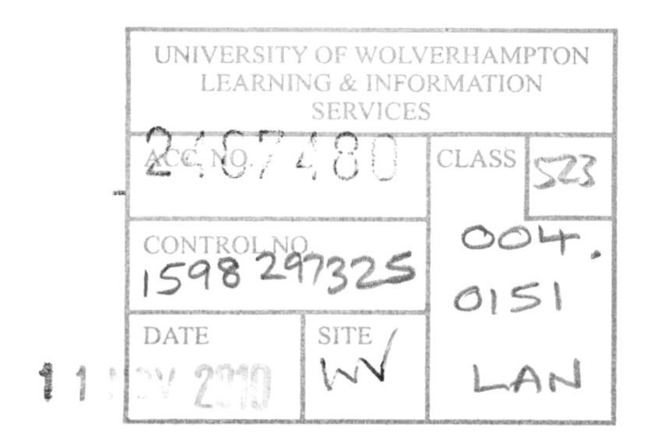

SYNTHESIS LECTURES ON QUANTUM COMPUTING #2

MORGAN & CLAYPOOL PUBLISHERS

ABSTRACT

In this text we present a technical overview of the emerging field of quantum computation along with new research results by the authors. What distinguishes our presentation from that of others is our focus on the relationship between quantum computation and computer science. Specifically, our emphasis is on the computational model of quantum computing rather than on the engineering issues associated with its physical implementation. We adopt this approach for the same reason that a book on computer programming doesn't cover the theory and physical realization of semiconductors. Another distinguishing feature of this text is our detailed discussion of the circuit complexity of quantum algorithms.

To the extent possible we have presented the material in a form that is accessible to the computer scientist, but in many cases we retain the conventional physics notation so that the reader will also be able to consult the relevant quantum computing literature. Although we expect the reader to have a solid understanding of linear algebra, we do not assume a background in physics. This text is based on lectures given as short courses and invited presentations around the world, and it has been used as the primary text for a graduate course at George Mason University. In all these cases our challenge has been the same: how to present to a general audience a concise introduction to the algorithmic structure and applications of quantum computing on an extremely short period of time. The feedback from these courses and presentations has greatly aided in making our exposition of challenging concepts more accessible to a general audience.

KEYWORDS

Quantum Computing, Quantum Algorithms, Quantum Information, Computer Science, Grover's Algorithm, Shor's Algorithm, Quantum Fourier Transform, Circuit Complexity, Computational Geometry, Computer Graphics, The Hidden Sub-Group Problem, Cryptoanalysis.

(ML):

To my sweetheart Michelle, because of all her love and support

To my mother Angelina, because of all her teachings and encouragement

To my babies Oliver, Oscar, and Chanel because all their meows make life very enjoyable

(JU):

To Zak, Max, and Tessa

Contents

Preface

In this text we present a technical overview of the emerging field of quantum computation along with new research results by the authors. This coverage of the field is sufficient to prepare the reader for more comprehensive physics-based books such as that by Nielsen[37]. What distinguishes our presentation from that of others is our focus on the relationship between quantum computation and computer science. Specifically, our emphasis is on the computational model of quantum computing rather than on the engineering issues associated with its physical implementation. We adopt this approach for the same reason that a book on computer programming doesn't cover the theory and physical realization of semiconductors. Another distinguishing feature of this text is our detailed discussion of the circuit complexity of quantum algorithms.

To the extent possible we have presented the material in a form that is accessible to the computer scientist, but in many cases we retain the conventional physics notation so that the reader will also be able to consult the relevant quantum computing literature. Although we expect the reader to have a solid understanding of linear algebra, we do not assume a background in physics. This text is based on lectures given as short courses and invited presentations around the world, and it has been used as the primary text for a graduate course at George Mason University. In all these cases our challenge has been the same: how to present to a general audience a concise introduction to the algorithmic structure and applications of quantum computing on an extremely short period of time. The feedback from these courses and presentations has greatly aided in making our exposition of challenging concepts more accessible to a general audience.

The organization of the book is as follows: Chapter 1 introduces the properties of quantum computing model which appear to offer the potential to perform computations more efficiently than is possible within the classical Turing model of computation. Chapter 2 discusses the theoretical framework of quantum computation in greater detail. In particular, classical and quantum complexity classes are delineated, and important quantum-algorithmic building blocks are identified that provide theoretical performance advantages over classical alternatives. Chapter 3 introduces the fundamental principle of quantum amplitude amplification. This principle permits a brute-force exhaustive examination of the set of all possible solutions to a given algorithmic problem to be performed in time that is sublinear in the size of that set. This is clearly not possible for arbitrary problems within the classical framework. Chapter 4 transitions from the theory of quantum algorithmics to the application of that theory to practical computational problems. We specifically examine several practical applications within the area of computational geometry. In Chapter 5 we move from practical applications to a detailed analysis of the Quantum Fourier Transform (QFT), which is a fundamental result that offers a means for defeating the most widely used classical public-key cryptographic scheme. Chapter 6 examines a class of computational problems that seems to be linked in a fundamental way to the power of quantum computing. Chapter 7 examines the circuit complexity of quantum algorithms and considers the specific cases where quantum parallelism actually provides greater computational power. This analysis questions the traditional criteria to establish when quantum computing actually offers a more powerful computational solution. Conclusions are presented in Chapter 8.

Marco Lanzagorta and Jeffrey Uhlmann
ITT Corporation and University of Missouri-Columbia
November 2008

Acknowledgements

We are thankful to our extraordinary editor, Mike Morgan, for his continued encouragement, advice, and perseverance. We would also like to thank Dr Robert Rosenberg from the US Naval Research Laboratory in Washington DC for kindly agreeing to review the first draft of our lecture, and whose comments proved to be very insightful and useful towards the improvement of our book.

ML acknowledges the support he received from ITT Corporation with regards to his efforts in the area of quantum information science. In particular, ML is thankful to Robert Blakely, Chuck Edlund, Gladys Weller, Thomas Payne, Michael Maciolek, Les Aker, and the late John Urban.

ML is also thankful to his many colleagues at the US Naval Research Laboratory who have fostered his interest in advanced information technologies. In particular, ML is thankful to Dr Keye Martin for his support, encouragement, and all the interesting discussions regarding quantum stuff.

And finally, ML acknowledges George Mason University for entrusting him with the instruction of a variety of advanced graduate courses in the areas of quantum computation and quantum information.

JU acknowledges the support received by the University of Missouri-Columbia.

Introduction

An advance in the field of computer science may consist of a new algorithm that is significantly more efficient than the prior state-of-the-art or it may consist of a theoretical proof that the current state-of-the-art cannot be significantly improved. The latter case imposes a fundamental constraint on the size of problems that can be solved in a fixed amount of time on a given computer. The only way to increase the size of problems that can be solved is to increase the speed of the computer. Since the 1960s, semiconductor size (and, consequently, processing power) has roughly doubled every two years according to what is referred to as "Moore's Law." Despite the fact that this improvement has been consistent for several decades, it is clear that it cannot continue indefinitely because of fundamental physical limitations. Specifically, by the year 2020 the circuits will be so small that their behavior will be dominated by quantum effects, and by 2050, the circuits will reach the minimum scale at which information can be physically represented.

These observations motivate interest in the implications of quantum theory to the evolution of computing technology during the next few decades. For example, can circuits be made robust to the effects of quantum phenomena? Or can quantum phenomena itself be harnessed to perform computations? The exploitation of quantum phenomena to perform computation is referred to as *quantum computing*. However, if a quantum computer simply offers improved performance due to the increased speed of quantum-scale circuits, then it is of more interest to the computer engineer than to the computer scientist. After all, increasing the clock speed of a processor does not affect the computational complexity of algorithms executed on that processor. If the architecture of the computer is changed to include some number P of processors, however, then different algorithms may offer superior complexity in terms of the new variable P. For example, optimal parallel exploitation of the P processors may improve the best possible complexity for solving a particular problem from $O(N)$ to $O(N/P)$. Of course, it is not generally possible to achieve an $O(P)$ complexity reduction because not all algorithms can be decomposed into $O(P)$ independent steps that can be executed in parallel throughout the running time of the algorithm.

A more fundamental change of architecture might provide the capability to store and manipulate, e.g., using analog circuits, arbitrary real numbers, instead of a discrete set of symbols. Such an architecture can be shown to offer dramatically greater computational power than the classical Turing machine. This increased power derives from the fact that large amounts of information can be packed into the infinite precision of a single real value, which can be manipulated in parallel when unit-cost operations are applied to that real value. This is of course entirely hypothetical because it depends on the assumption that infinite precision can be maintained during those operations, and there is no reason to believe that such an architecture is physically realizable. A quantum computer similarly exploits the maintenance of real – in fact, complex – values, but its power doesn't exploit an assumption of infinite precisions. Rather, it exploits a property of quantum mechanics that allows special statistical relationships among a set of states to be manipulated in parallel.

So-called "quantum parallelism" is the key property that permits – at least in theory – a quantum computer to support algorithmic solutions that are provably more efficient than what is possible for any classical algorithm. There are no doubts about the correctness of the underlying theory of quantum mechanics, but there are significant issues relating to the physical realization of that theory in hardware. As will be discussed, these issues are not purely practical: there are theoretical subtleties associated with the scaling of quantum resources (in terms of numbers of gates) necessary to exploit quantum parallelism. More specifically, it is possible that the scaling of the number of quantum gates necessary to implement certain quantum algorithms exactly balances the apparent quantum parallelism such that the net result is equivalent to classical parallelism.

CHAPTER 1

The Algorithmic Structure of Quantum Computing

Quantum Computing (QC) has become a very active area of research in recent years. This interest is motivated by the recognition that certain types of quantum phenomena behave in ways that cannot be simulated efficiently using ordinary computers [17]. Furthermore, problems whose solutions can be expressed in terms of one of these phenomena can be solved more efficiently using the quantum phenomenon itself as a computational tool.

This way of thinking about computation is analogous to many familiar techniques for simulating natural phenomena, such as, the effects of earthquakes on structures. For example, it is possible to simulate on a computer the wooden frame of a house, the walls, the ceilings, the roof, etc., and then compute what would happen if shaken by a magnitude seven earthquake. A challenge arises from the high-fidelity modeling of all the details of a real house. If every splinter of wood has to be represented, the amount of computer memory and simulation time will become impractically large. Therefore, it makes sense to simply build a house on a platform that can be made to shake appropriately and then observe the effect on the structure. In other words, modeling the physical situation and letting it evolve naturally provides a direct computational mechanism for answering the question of how a magnitude seven earthquake will affect the structure.

The analogy of quantum computing with a physical simulation – as opposed to a virtual simulation on a computer – is not quite exact because in theory all classical physics can be simulated efficiently using classical computers. This is because all of the variables in a classical system can be represented with variables in a software simulation. It may be difficult and tedious to create the high-fidelity software simulation, but it can certainly be done in theory. Quantum systems, however, evolve in a state space that is exponentially larger than the number of parameters required to define a particular physical state. Therefore, a simulation of a quantum system on a classical computer would have to explicitly represent this exponentially larger state space, whereas, the evolution of the system itself does not.

The recognition that quantum phenomena can be exploited to compute results that cannot be efficiently computed on a classical computer is exciting because it indicates a more general and powerful model of computation than exists within the classical framework. The challenge then is to understand the characteristics of quantum phenomena that provide this computational advantage and determine how to effectively apply them to solve practical computational problems. Fortunately, it turns out that a straighforward generalization of the binary logic gates used in classical computers is sufficient to capture the full power of quantum computing. In this case, quantum gates operate on quantum states. This framework for quantum computation is often called the *quantum circuit model*[1].

1.1 UNDERSTANDING QUANTUM ALGORITHMICS

Most introductions to quantum computing are *physics-centric* in the sense that they focus first on theoretical quantum physics and how quantum dynamical systems can be used to perform computations that cannot be performed efficiently on a classical computer [40, 47, 13]. Such a presentation may mirror the historical way in which quantum computing emerged from the evolving theory of quantum physics, but it unnecessarily obscures the abstract computational model implied by quantum theory with the details of quantum mechanics as they apply to physical systems. This kind of approach is analogous to introducing Boolean algebra in terms of transistors or Java programming in terms of the behavior of electrons in semiconductors.

The objective of this chapter is to introduce quantum computing as an abstract mathematical framework stripped of any connection to the physical world. The only compromise in this regard is the use of notation and terminology that is inherited from quantum physics. It would be possible to replace physics terms, such as, *measurement* with more conventional computer science terms, such as, *read operation*, but to do so would effectively render the bulk of the physics-oriented quantum computing literature inaccessible.

In what follows we will enumerate and discuss eight properties that describe the quantum computational model. These properties are not independent of each other, so they do not constitute an axiomatic set from which the quantum model is formally defined, but they each capture an aspect of the quantum computing model that stands in stark contrast to the classical computing model.

1.1.1 QUANTUM COMPUTING PROPERTY #1

The first property of the quantum computing model is actually a definition of what is meant by information in the quantum world. Specifically, the qubit *is defined as a generalization of the classical* bit *as the new unit of (quantum) information.*

A classical bit is a scalar variable which has a single value of either 0 or 1. The bit's value is unique, deterministic, and unambigous. On the other hand, a qubit is more general in the sense that it represents a state defined by a pair of complex numbers, (a, b), which together express the probability that a reading of the value of the qubit will give a value of 0 or 1. Thus, a qubit can be in the state of 0, 1, or some mixture - referred to as a *superposition* - of the 0 and 1 states. The weights of 0 and 1 in this superposition are determined by (a, b) in the following way:

$$qubit = (a, b) = a \cdot 0_{bit} + b \cdot 1_{bit} \; . \tag{1.1}$$

[1]It is important to note that the circuit model of quantum computation is not the only quantum model of computation. To date, several other models, such as, adiabatic, topological, and one-way quantum computing have been proposed. Some of them have been shown to be computationally equivalent. That is, they can solve the exact same computational problems. However, a detailed discussion of these alternate models of computation is beyond the scope of this introduction-level lecture.

Clearly, in strong contrast to a bit, the state of a qubit can be both 0 and 1 simultaneously. Stated more generally, the state of a qubit is represented as a linear combination of the 0 state and the 1 state, where the weights are defined by the values a and b. A bit can, therefore, be thought of as a special case of a qubit in which there is no superposition. For example, the qubit $(1, 0)$ represents a bit in the state 0, while the qubit $(0, 1)$ is equivalent to a bit in the state 1.

Bra-ket is a concise notation originally developed by the physicist Paul M. Dirac, and it is conventionally used to express quantum states [14]. It is a generalization of common vector notation in which $\langle\psi|$ is a row vector (read as "bra psi") and $|\boldsymbol{\psi}\rangle$ is a complex conjugate column vector (read as "ket psi"), and the inner product $\langle\boldsymbol{\psi}|\boldsymbol{\phi}\rangle$ is referred to as a *bracket* (which is the origin of the root terms *bra* and *ket*). In this notation the state of a single qubit can be written as:

$$
\begin{aligned}
|\boldsymbol{\Psi}\rangle &= a|\mathbf{0}\rangle + b|\mathbf{1}\rangle \\
&\text{or} \\
\langle\Psi| &= a^*\langle\mathbf{0}| + b^*\langle\mathbf{1}| \,.
\end{aligned}
$$

It should be noted that $|\mathbf{0}\rangle$ is not a zero vector; rather the zero is just a label for a unit basis vector which is orthogonal to the other basis vector $|\mathbf{1}\rangle$. In other words, the state of a qubit spans a complex two-dimensional space, and the choice of orthogonal basis vectors corresponding to states 0 and 1 is arbitrary. For instance, in the most commonly used representation, called the *computational basis*, we have:

$$
\begin{aligned}
\langle\mathbf{0}| &= (1, 0) \\
\langle\mathbf{1}| &= (0, 1)
\end{aligned}
$$

and:

$$
\begin{aligned}
|\mathbf{0}\rangle &= \begin{pmatrix} 1 \\ 0 \end{pmatrix} \\
|\mathbf{1}\rangle &= \begin{pmatrix} 0 \\ 1 \end{pmatrix} .
\end{aligned}
$$

Because $|\mathbf{0}\rangle$ and $|\mathbf{1}\rangle$ are orthogonal unit vectors:

$$
\langle\mathbf{0}|\mathbf{0}\rangle = \langle\mathbf{1}|\mathbf{1}\rangle = 1 \tag{1.2}
$$

and

$$
\langle\mathbf{0}|\mathbf{1}\rangle = \langle\mathbf{1}|\mathbf{0}\rangle = 0 \,. \tag{1.3}
$$

Let us remark, once more, that $|\mathbf{0}\rangle$ and $|\mathbf{1}\rangle$ represent logical bits of classical information (0 and 1, respectively). Therefore, a qubit can be understood as a complex linear superposition of classical bits.

Alternatively, a qubit can be understood as a complex vector in an abstract classical bit space. Indeed, in vector notation the linear combination of states for a single qubit, $a|\mathbf{0}\rangle + b|\mathbf{1}\rangle$, would be expressed as:

$$
\vec{v} = a\vec{\imath} + b\vec{\jmath} \tag{1.4}
$$

with

$$\vec{\imath} \cdot \vec{\imath} = \vec{\jmath} \cdot \vec{\jmath} = 1 \tag{1.5}$$
$$\vec{\imath} \cdot \vec{\jmath} = \vec{\jmath} \cdot \vec{\imath} = 0\,. \tag{1.6}$$

And $\vec{\imath}$ is a vector parallel to $|\mathbf{0}\rangle$ and $\vec{\jmath}$ is a vector parallel to $|\mathbf{1}\rangle$. Clearly, both representations are very similar. However, the advantage of bra-ket notation over conventional vector notation becomes evident when we generalize to multiple qubits. Whereas a single qubit represents a linear combination of two states, a superposition of n-qubits represents a linear combination of 2^n states. For example, a 2-qubit state would be a linear combination of four basis states represented by 2 bit digits:

$$|\boldsymbol{q}^{(2)}\rangle = \alpha|\mathbf{00}\rangle + \beta|\mathbf{01}\rangle + \gamma|\mathbf{10}\rangle + \delta|\mathbf{11}\rangle \tag{1.7}$$

which is equivalent to a four-dimensional vector in an abastract space spanned by orthonormal 2-bit states. As before, the superposition coefficients α, β, γ, and δ are complex numbers.

This expression results from the fact that a 2-qubit state can be constructed from the tensor product of two 1-qubit states:

$$|\boldsymbol{q}^{(2)}\rangle = (a|\mathbf{0}\rangle + b|\mathbf{1}\rangle) \otimes (c|\mathbf{0}\rangle + d|\mathbf{1}\rangle) \tag{1.8}$$
$$= ac|\mathbf{0}\rangle \otimes |\mathbf{0}\rangle + ad|\mathbf{0}\rangle \otimes |\mathbf{1}\rangle + bc|\mathbf{1}\rangle \otimes |\mathbf{0}\rangle + bd|\mathbf{1}\rangle \otimes |\mathbf{1}\rangle \tag{1.9}$$

and now we can define an extended computational basis to accomodate these 2-qubit states as:

$$|\mathbf{00}\rangle = |\mathbf{0}\rangle \otimes |\mathbf{0}\rangle \tag{1.10}$$
$$|\mathbf{01}\rangle = |\mathbf{0}\rangle \otimes |\mathbf{1}\rangle \tag{1.11}$$
$$|\mathbf{10}\rangle = |\mathbf{1}\rangle \otimes |\mathbf{0}\rangle \tag{1.12}$$
$$|\mathbf{11}\rangle = |\mathbf{1}\rangle \otimes |\mathbf{1}\rangle\,. \tag{1.13}$$

And then

$$|\boldsymbol{q}^{(2)}\rangle = ac|\mathbf{00}\rangle + ad|\mathbf{01}\rangle + bc|\mathbf{10}\rangle + bd|\mathbf{11}\rangle \tag{1.14}$$

It is important to note that, in general, we can have 2-qubit states which cannot be represented as the tensor product of two qubits. For instance:

$$|\boldsymbol{q}^{(2)}\rangle = \frac{1}{\sqrt{2}}(|\mathbf{00}\rangle + |\mathbf{11}\rangle) \neq |\boldsymbol{q}_a^{(1)}\rangle \otimes |\boldsymbol{q}_b^{(1)}\rangle\,. \tag{1.15}$$

These types of quantum superpositions, which cannot be written as the tensor product of more elementary states, are said to be *entangled*.

In any event, we can follow this same procedure described above to build the bases for the qubits of arbitrary size. Most generally, the state Ψ of a quantum register of n-qubits is expressed as:

$$|\boldsymbol{\Psi}\rangle = \sum_{i=0}^{2^n-1} \alpha_i |\boldsymbol{i}\rangle \tag{1.16}$$

where each i corresponds to a distinct n-bit binary vector in the superposition of 2^n states. That is, $|\boldsymbol{i}\rangle$ is an element of the computational basis for n-qubits, and i is a binary enumeration of the elements of this basis:

$$|\Psi\rangle = \alpha_0|\mathbf{00...00}\rangle + \alpha_1|\mathbf{00...01}\rangle + ... + \alpha_{N-1}|\mathbf{11...11}\rangle \tag{1.17}$$

where $N = 2^n$, a convention that we will consistently follow across the entire text.

In the same way as before, we can define the 2^n-dimensional orthonormal vector representations of these n-qubits as:

$$\begin{aligned}
\langle \mathbf{00...0}| &= (1, 0, 0, ..., 0) && (1.18)\\
\langle \mathbf{00...1}| &= (0, 1, 0, ..., 0) && (1.19)\\
... &\quad ... && (1.20)\\
\langle \mathbf{11...1}| &= (0, 0, 0, ..., 1)\ . && (1.21)
\end{aligned}$$

Although it has not been mentioned here before, the weights (α_i) in the linear combination are taken to have a Euclidean norm of unity:

$$\begin{aligned}
\langle \Psi|\Psi\rangle &= \sum_{i=0}^{2^n-1} \alpha_i \alpha_i^* && (1.22)\\
&= \sum_{i=0}^{2^n-1} |\alpha_i|^2 && (1.23)\\
&= 1\ . && (1.24)
\end{aligned}$$

The reason for such a choice will be clear when we discuss the second property of the quantum computational model.

Before concluding with the discussion of Property #1, it is important to state that the computational basis is not the only basis to operate quantum states in the quantum computing model. In the case of 2-qubits, for instance, we could use the alternative basis which results from a rotation of the computational basis. The elements of this basis are:

$$\begin{aligned}
|+\rangle &= \frac{|\mathbf{0}\rangle + |\mathbf{1}\rangle}{\sqrt{2}} && (1.25)\\
|-\rangle &= \frac{|\mathbf{0}\rangle - |\mathbf{1}\rangle}{\sqrt{2}}\ . && (1.26)
\end{aligned}$$

In general, any other orthonormal basis will be equally viable. The decision of what basis to use depends on the quantum developer, as in some situations certain basis offer a simplified structure of the algorithm. As long as the basis are used on a consistent manner, it is perfectly correct to use any of them.

1.1.2 QUANTUM COMPUTING PROPERTY #2

Quantum computing is a probabilistic computational model.

When an n-qubit quantum register contains a superposition of 2^n states, the application of a read operation to the register will cause the superposition to "collapse" to a single classical state. That is, a measurement of a 2-qubit state produces a 2-bit result. The specific state to which it collapses is probabilistic with statistics determined by the weights in the linear combination. For example, a read - or *measurement* - of a 2-qubit register R given by

$$|\boldsymbol{R}\rangle = \alpha|\mathbf{00}\rangle + \beta|\mathbf{01}\rangle + \gamma|\mathbf{10}\rangle + \delta|\mathbf{11}\rangle \tag{1.27}$$

will obtain the classical bit state 00 with probability $|\alpha|^2$, the classical bit state 01 with probability $|\beta|^2$, and so on.

In general, measurement of a n-qubit state produces n-bits of classical information, and the probability to obtain the state i is given by:

$$P_i = |\langle \boldsymbol{i}|\boldsymbol{R}\rangle|^2 \tag{1.28}$$

where $\langle \mathbf{i}|$ is an n-bit binary vector in the computational basis.

As with most probabilistic problems, we want to impose the condition that the sum of the probabilities for all possible outsomes should equal 1:

$$P_{total} = \sum_i P_i = 1 \; . \tag{1.29}$$

In the previous example this means that:

$$|\alpha|^2 + |\beta|^2 + |\gamma|^2 + |\delta|^2 = 1 \; . \tag{1.30}$$

Therefore, we need to impose the condition that quantum registers should always be normalized to unity.

It is critical to understand that once a measurement is applied to obtain a state, all subsequent measurements will obtain that same state. Thus, if the state 10 is read from $|\boldsymbol{R}\rangle$, the superposition given above has collapsed so that $\gamma = 1$ and all other weights are zero. At this point R is equivalent to a classical register containing the state 10:

$$|\boldsymbol{R}\rangle \rightarrow |\mathbf{10}\rangle \; . \tag{1.31}$$

An operation applied to a classical register changes its content from one bit vector to another bit vector. In the above example of the quantum register R, after a measurement yields the state 01, a classical operation which transforms the state to 11 can be interpreted as setting $\delta = 1$ and $\gamma = 0$. Now a measurement of the register obtains the state 11 with probability 1.

In general, any classical operation can be represented as a permutation matrix applied to the vector of α_i weights, or equivalently, as a permutation of the elements of each state vector in the superposition. As should be expected, the normalization of the weights is preserved after a permutation transformation. For example:

$$\alpha|\mathbf{00}\rangle + \beta|\mathbf{01}\rangle + \gamma|\mathbf{10}\rangle + \delta|\mathbf{11}\rangle \longrightarrow \gamma|\mathbf{00}\rangle + \delta|\mathbf{01}\rangle + \alpha|\mathbf{10}\rangle + \beta|\mathbf{11}\rangle \tag{1.32}$$

with:

$$|\alpha|^2 + |\beta|^2 + |\gamma|^2 + |\delta|^2 = |\gamma|^2 + |\delta|^2 + |\alpha|^2 + |\beta|^2 = 1 \,. \tag{1.33}$$

The QC framework generalizes the class of possible operations to include any unitary transformation. Let us recall that a unitary transformation U satisfies:

$$U^{-1} = U^{\dagger} \tag{1.34}$$

and therefore:

$$U^{\dagger}U = I \tag{1.35}$$

where I is the identity matrix and $U^{\dagger}$ is the complex conjugate transpose matrix of U. So, the transformation of a vector v as:

$$v \rightarrow w = Uv \tag{1.36}$$

will preserve the magnitude of v because:

$$|w|^2 = w^{\dagger}w = v^{\dagger}U^{\dagger}Uv = v^{\dagger}v = |v|^2 \,. \tag{1.37}$$

In the quantum context this means that the Euclidean norm of the states/weights is preserved to be unity under a unitary transformation U. If:

$$U \begin{pmatrix} \alpha \\ \beta \\ \gamma \\ \delta \end{pmatrix} \rightarrow \begin{pmatrix} \alpha' \\ \beta' \\ \delta' \\ \gamma' \end{pmatrix} \tag{1.38}$$

then U affects the respective probabilities associated with each state as:

$$\alpha|\mathbf{00}\rangle + \beta|\mathbf{01}\rangle + \gamma|\mathbf{10}\rangle + \delta|\mathbf{11}\rangle \longrightarrow \alpha'|\mathbf{00}\rangle + \beta'|\mathbf{01}\rangle + \gamma'|\mathbf{10}\rangle + \delta'|\mathbf{11}\rangle \tag{1.39}$$

with:

$$|\alpha|^2 + |\beta|^2 + |\gamma|^2 + |\delta|^2 = |\alpha'|^2 + |\beta'|^2 + |\gamma'|^2 + |\delta'|^2 = 1 \,. \tag{1.40}$$

This is an expected feature of a well defined probabilistic computing model. Indeed, if the quantum computational model is a probabilistic computational model, then the probabilities should always add up to one. And this has to be true even after performing transformations to the states. As a consequence, except for measurements, unitary operations are the only type of transformations allowed in QC.

Of course, neither classical nor quantum operations would ever be explicitly implemented using a $2^n \times 2^n$ matrix because this would lead to quantum circuits of exponential size and/or exponentially large computational times. Thus, effective quantum algorithms should have computational operations that correspond to highly-structured unitary matrices which can be represented implicitly with low complexity, e.g., using a small number of logic gates, proportional to n [2]. The reason for abstracting computation to linear algebra in a 2^n-dimensional space is that it permits the computational power of QC over CC to be examined in its complete generality rather than in terms of particular operations.

Later chapters will discuss particular transformations that will form the building blocks for new quantum algorithms which have better complexity than the best classical alternative. A distinguishing feature of these quantum algorithms is that each will have a fixed probability of producing a correct result. Therefore, a constant number of executions of the algorithm is sufficient to ensure that a solution is found with probability arbitrarily close to unity.

1.1.3 QUANTUM COMPUTING PROPERTY #3

Measurements (read operations) in the quantum computational model are destructive.

As has been discussed, reading the content of a quantum register yields a classical state. That means that any other states that were in a superposition with the measured state are lost. Furthermore, there is no way to recover these lost states after a superposition is collapsed due to a measurement. That is, measurements are destructives and there is no operation to reverse their effect.

The importance of Property #3 cannot be overstated because it implies that although an n-qubit quantum register can contain a superposition of states of size 2^n (the number of states that can be represented with n-bits), the end result can be only one measured state with n-bits of logical information.

If we have a multi-qubit register, it is possible to only measure the state of specific qubits. However, the state of the register after measurement will depend on its entanglement. For instance, suppose we have two qubits a and b in a non-entangled state:

$$|\mathbf{q_{ab}}\rangle = (\alpha|\mathbf{0_a}\rangle + \beta|\mathbf{1_a}\rangle) \otimes (\gamma|\mathbf{0_b}\rangle + \delta|\mathbf{1_b}\rangle) \tag{1.41}$$

If we only measure the state of qubit a, we will find it in the state 0 with probability $|\alpha|^2$ and the register will collapse to:

$$|\mathbf{q_{ab}}\rangle = |\mathbf{0_a}\rangle \otimes (\gamma|\mathbf{0_b}\rangle + \delta|\mathbf{1_b}\rangle) \tag{1.42}$$

and similarly if we find a in the state 1 with probability $|\beta|^2$. In both cases, the measurement of qubit a does not disturb the state of qubit b. As such, a subsequent measurement of qubit b is completely independent of the measurement of qubit a.

[2]The circuit complexity of quantum algorithms is further discussed in chapter 7.

On the other hand, let us suppose that we have a two qubit quantum register in an entangled state given by:

$$|Q\rangle_{ab} = \frac{1}{\sqrt{2}}\left(|\mathbf{0}_a\mathbf{0}_b\rangle + |\mathbf{1}_a\mathbf{1}_b\rangle\right) \tag{1.43}$$

If we measure a, we will find it in the state 0 with probability $1/2$ and the register will collapse to:

$$|Q\rangle_{ab} = |\mathbf{0}_a\mathbf{0}_b\rangle \tag{1.44}$$

and then any subsequent measurement of b will find it to be in 0 with total certainty. Similarly, if we had measured a in the state 1, then a subsequent measurement of b will find it in 1 with total certainty. Therefore, the measured state of b depends on the previous measurement of a. As a consequence, when dealing with entangled states, the measurements of a qubit will affect the state of other qubits.

Furthermore, Property #3 eliminates the ability to check the value of the quantum register during the execution of a quantum algorithm. That means that there is no way to obtain and use intermediate results or perform other types of operations that are common for classical algorithms. For instance, debugging print statements are not allowed in the quantum model.

This irreversible loss of information is one of the challenges for effective design of quantum algorithms. However, it is important to note that the power of quantum computing derives from its ability to manipulate the superposition to increase the probability of measuring a desired solution state. Therefore, reading the content of the quantum register must be the last step of the quantum algorithm.

1.1.4 QUANTUM COMPUTING PROPERTY #4

Quantum computing operates in a computational space that is exponentially larger than what is possible with classical registers.

Let us remember that a single bit can only store a single memory address, either 0 or 1. On the other hand, a single qubit can simultaneously store a mixture of the two states, i.e., a measurement of the qubit will obtain 0 with probability p and will obtain 1 with with probability $1 - p$.

In general, a classical n-bit register can index $N = 2^n$ states, but the index of only one state can be stored and transformed in the register. A quantum register, on the other hand, can store a superposition of indices to all N states, and the weights on all N states can be transformed in parallel.

In particular, the classical storage of indices to N unique states requires $N\log(N)$ bits. For example, 24 bits can be used to store 8 unique 3-bit addresses. On the other hand, quantum starage of N indices requires $Log(N)$ qubits. That is, a 24-qubit quantum register can store a superposition of more than $2^{24} \approx 16$ million distinct addresses. Or alternatively, 3 qubits are enough to store 8 unique 3-bit addresses.

As will be shown in a subsequent chapter, this can allow an array of size N to be searched using quantum parallelism to obtain the index of any desired element after fewer than N sequential steps. In other words, a quantum algorithm can truly examine N elements of an array in parallel through a superposition of indices.

Property #4 encompasses one of the most important advantages of the quantum model. However, as we already mentioned, upon measurement of a n-qubit register containing a superposition of 2^n distinct addresses, we can only extract the address expressed by a string of n-bits. That is, the quantum model has an exponentially large computational space, but most of this space is inaccessible to us once we perform a measurement. The challenge for developing a quantum algorithm is to exploit the huge computational space before a measurement is performed.

1.1.5 QUANTUM COMPUTING PROPERTY #5

Except for measurements, all operations on qubits must be reversible.

Only reversible operations can be applied to a quantum register without causing the superposition to collapse. This is a consequence of the more general condition that a quantum superposition can only be preserved under unitary transformations. Because a unitary matrix U satisfies $U^{-1} = U^{\dagger}$, it can be reversed to retrieve the original state of the system. That is, there is an inverse operation:

$$|\Psi\rangle \rightarrow U|\Psi\rangle = |\Psi'\rangle \tag{1.45}$$

$$|\Psi'\rangle \rightarrow U^{-1}|\Psi'\rangle = |\Psi\rangle \; . \tag{1.46}$$

Actually, reversibility can be thought of as the reason why only unitary operators preserve the superposition: any irreversible operation results in a loss of information, and is essentially a measurement and thus causes a collapse of the superposition. In a sense, Property #5 is a consequence of Property #2, which states that the quantum computational model is probabilistic. As discussed before, conservation of the total probability can only be achieved with unitary operations.

Let us consider one of the simplest quantum operators, the Control-Not operation (CNOT). The CNOT operator involves a control bit and a target bit. If the control bit is 0, then the operator does nothing to the target bit. But if the control bit is 1, then it negates the state of the target bit. The truth table of the CNOT operation then looks like:

a	b	a'	b'
0	0	0	0
0	1	0	1
1	0	1	1
1	1	1	0

where a is the control bit, and b the target bit. This operation is clearly reversible; as for each output pair (a', b'), we can detemine without ambiguity the input pair (a, b). This is true in general, that is, 1 to 1 bijective binary functions always represent reversible operations.

We can easily generalize the CNOT for the quantum domain. To this end we can use the computational basis of 2-qubit states to build a quantum operator that implements the CNOT truth table. Thus, the effect of a CNOT on a general 2-qubit quantum register looks like:

$$\begin{aligned} |\boldsymbol{R}\rangle &= \alpha|\mathbf{00}\rangle + \beta|\mathbf{01}\rangle + \gamma|\mathbf{10}\rangle + \delta|\mathbf{11}\rangle && (1.47) \\ CNOT|\boldsymbol{R}\rangle &= \alpha|\mathbf{00}\rangle + \beta|\mathbf{01}\rangle + \gamma|\mathbf{11}\rangle + \delta|\mathbf{10}\rangle \,. && (1.48) \end{aligned}$$

Observe how the effect of the CNOT gate is implemented on each and every single state of the superposition matching the values found in the truth table. We can also write CNOT in matrix form as follows:

$$CNOT = \begin{pmatrix} 1 & 0 & 0 & 0 \\ 0 & 1 & 0 & 0 \\ 0 & 0 & 0 & 1 \\ 0 & 0 & 1 & 0 \end{pmatrix} . \tag{1.49}$$

And therefore:

$$CNOT|\boldsymbol{R}\rangle = \begin{pmatrix} 1 & 0 & 0 & 0 \\ 0 & 1 & 0 & 0 \\ 0 & 0 & 0 & 1 \\ 0 & 0 & 1 & 0 \end{pmatrix} \begin{pmatrix} \alpha \\ \beta \\ \gamma \\ \delta \end{pmatrix} = \begin{pmatrix} \alpha \\ \beta \\ \delta \\ \gamma \end{pmatrix} \tag{1.50}$$

which are clearly equivalent.

The diagramatic representation of the CNOT gate is shown if Figure 1.1. Here the two input states are:

$$\begin{aligned} |\boldsymbol{\Psi_a}\rangle &= \alpha|\mathbf{0}\rangle + \beta|\mathbf{1}\rangle && (1.51) \\ |\boldsymbol{\Psi_b}\rangle &= \gamma|\mathbf{0}\rangle + \delta|\mathbf{1}\rangle && (1.52) \\ & && (1.53) \end{aligned}$$

which through the tensor product form the input state:

$$\begin{aligned} |\boldsymbol{\Psi_i}\rangle &= |\boldsymbol{\Psi_a}\rangle \otimes |\boldsymbol{\Psi_b}\rangle && (1.54) \\ &= (\alpha|\mathbf{0}\rangle + \beta|\mathbf{1}\rangle) \otimes (\gamma|\mathbf{0}\rangle + \delta|\mathbf{1}\rangle) && (1.55) \\ &= \alpha\gamma|\mathbf{00}\rangle + \alpha\delta|\mathbf{01}\rangle + \beta\gamma|\mathbf{10}\rangle + \beta\delta|\mathbf{11}\rangle && (1.56) \end{aligned}$$

and the final state is given by:

$$|\boldsymbol{\Psi_f}\rangle = \alpha\gamma|\mathbf{00}\rangle + \alpha\delta|\mathbf{01}\rangle + \beta\gamma|\mathbf{11}\rangle + \beta\delta|\mathbf{10}\rangle \,. \tag{1.57}$$

An equivalent statement of Property #5 is that any loss of information resulting from a transformation of a quantum superposition constitutes a measurement. Fortunately, any non-reversible classical logic gate/circuit can be simulated using reversible gates. For example, consider binary addition (also known as exclusive-or), which has the following truth table:

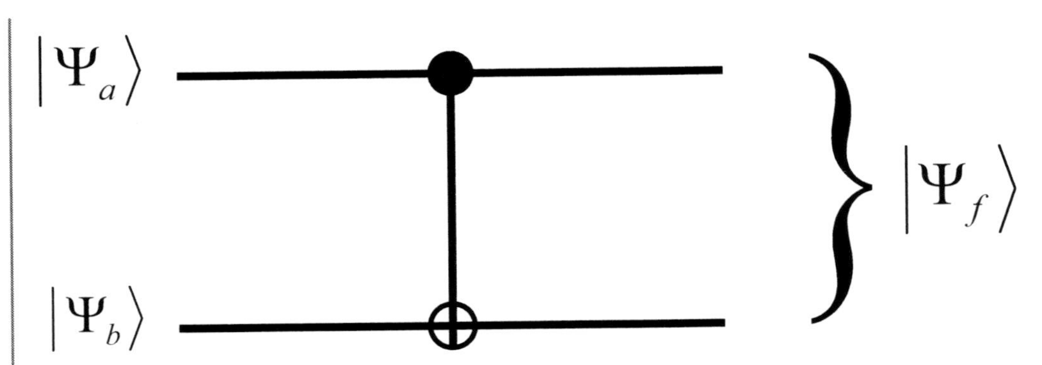

Figure 1.1: Diagramatic representation of the CNOT gate. The solid dot represents the connection to a control bit. The cross represents the effect of a NOT operation on a target bit only when the control bit allows it.

a	b	a + b
0	0	0
0	1	1
1	0	1
1	1	0

We can easily observe that binary addition is non-reversible because the values of a and b cannot be uniquely determined from the result $a + b$. However, the operation can be made reversible by augmenting it with an extra variable c:

a	b	c	a	b	a + b + c
0	0	0	0	0	0
1	0	0	1	0	1
0	1	0	0	1	1
1	1	0	1	1	0
0	0	1	0	0	1
1	0	1	1	0	0
0	1	1	0	1	0
1	1	1	1	1	1

Clearly, this is a reversible operation because given a set of output numbers $(a, b, a + b + c)$, we can always determine the value of the input (a, b, c). Furthermore, by choosing $c = 0$ we get the same functionality as the binary addition of a and b [3].

[3]It is important to note that this modification to the original operation is not unique.

We can use the above table to build a quantum unitary operator, $U(+)$, which can be used in place of binary addition because the value in the variable c permits the values of a and b to be recovered.

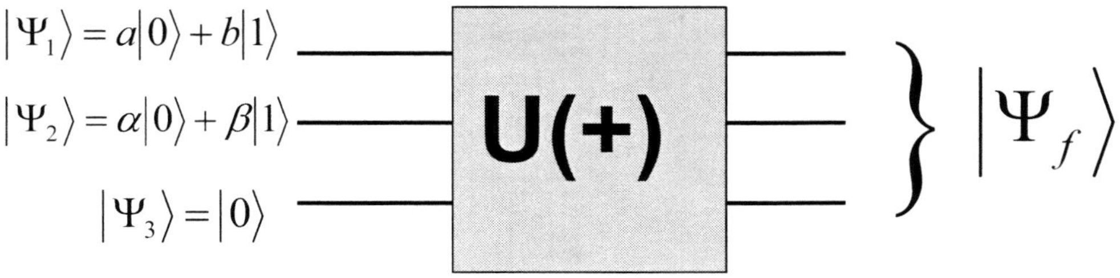

Figure 1.2: Reversible implementation of the binary addition on quantum states.

A reversible quantum gate can now be constructed that implements binary addition on two qubits. The diagramatic representation of the quantum circuit that implements this gate is shown in Figure 1.2.

The computational analysis of quantum circuits such as this can be done as follows. Before each operation we trace a vertical line. And for each of these vertical lines we write the state of the system.

In the binary addition example we have only one operator in the quantum circuit. This means that the circuit analysis only involves two vertical lines, one before and one after the operator, corresponding to the initial and final states of the system.

For the initial state we have the tensor product of the 3 input states:

$$\begin{aligned} |\psi_i\rangle &= |\psi_1\rangle \otimes |\psi_2\rangle \otimes |\psi_3\rangle \\ &= (a|0\rangle + b|1\rangle) \otimes (\alpha|0\rangle + \beta|1\rangle) \otimes |0\rangle \\ &= a\alpha|000\rangle + a\beta|010\rangle + b\alpha|100\rangle + b\beta|110\rangle \ . \end{aligned}$$

Now, the effect of the operator is determined by its logical table, which acts on each element of the superposition:

$$|a, b, c\rangle \rightarrow U(+)|a, b, c\rangle = |a, b, a+b+c\rangle \ . \tag{1.58}$$

So, for instance,

$$|010\rangle \rightarrow U(+)|010\rangle = |011\rangle \ . \tag{1.59}$$

Therefore, the final state is given by:

$$\begin{aligned} |\psi_f\rangle &= U(+)|\psi_i\rangle \\ &= a\alpha|000\rangle + a\beta|011\rangle + b\alpha|101\rangle + b\beta|110\rangle \ . \end{aligned}$$

Let us note that even though $U(+)$ is a quantum operator that implements binary addition, the additions are performed with the bits representing each state of the superposition. That is, the final result is different from:

$$|\psi_f\rangle \neq |\psi_1\rangle + |\psi_2\rangle + |\psi_3\rangle \,. \tag{1.60}$$

As indeed, such an operation would violate unitarity.

Furthermore, observe that the 4 possible values of the binary addition of two 1-bit elements have been computed in a single computational step. This important feature of quantum computing will be discussed as Property #6 of the quantum computing model.

In general, all classical logic circuits can be augmented in a similar fashion to obtain reversible substitutes at a cost of some extra gates. Furthermore, it can be shown that the change from an irreversible to a reversible circuit can be efficiently accomplished with only a constant overhead.

The algorithmic relevance of Property #5 is clear: except for measurements, all the operations required by the algorithm have to be reversible, represented by unitary operators.

1.1.6 QUANTUM COMPUTING PROPERTY #6

The Quantum Computing model offers instrinsic parallelism.

The fact that a unitary transformation U can be applied to simultaneously transform all states in a superposition has already been discussed. For example, a given U can be applied to $|\boldsymbol{R}\rangle$,

$$|\boldsymbol{R}\rangle = a|\mathbf{00}\rangle + b|\mathbf{01}\rangle + c|\mathbf{10}\rangle + d|\mathbf{11}\rangle \tag{1.61}$$

as:

$$U|\boldsymbol{R}\rangle = aU|\mathbf{00}\rangle + bU|\mathbf{01}\rangle + cU|\mathbf{10}\rangle + dU|\mathbf{11}\rangle \tag{1.62}$$

and all four state transformations are performed in a single computational step.

Clearly, Property #6 represents one of the great advantages of the quantum computational model. For instance, we can compute the value of a binary function for all N possible values of its input variable in a single computational step. Indeed, suppose that f is a binary function:

$$f : \{0, 1\}^n \to \{0, 1\} \tag{1.63}$$

and also suppose that U is a reversible quantum operator that performs the following operation:

$$U|\boldsymbol{x}\rangle|\mathbf{0}\rangle = |\boldsymbol{x}\rangle|\boldsymbol{f}(\boldsymbol{x})\rangle \tag{1.64}$$

where x is a binary number which enumerates the elements of the n-qubit computational basis. Then, if we start from a uniform quantum superposition to which we apply the operator U:

$$\frac{1}{\sqrt{N}} \sum_{x=0}^{N-1} |\boldsymbol{x}\rangle|\mathbf{0}\rangle \xrightarrow{U} \frac{1}{\sqrt{N}} \sum_{x=0}^{N-1} |\boldsymbol{x}\rangle|\boldsymbol{f}(\boldsymbol{x})\rangle \tag{1.65}$$

we have computed all the values of f in a single computational step. The classical complexity to perform the N evaluations of f is clearly $\Theta(N)$. This example clearly demonstrates the power of QC over CC in this instance.

However, let us recall that if we try to read the register after the evaluation of f has been completed, we will get obtain something like:

$$|\boldsymbol{x_i}\rangle|\boldsymbol{f(x_i)}\rangle \tag{1.66}$$

with probability $1/N$, and the superposition with all the function values is destroyed.

Of course, one could try to apply this process in a sequence of evaluations and measurements. However, to extract all N values would require $O(N\log(N))$ repetitions. Clearly, this method is suboptimal because it can be performed classically in only $\Theta(N)$ time. Thus, while this is an illustrative example, it is not a practical one. In subsequent chapters we will discuss in detail practical applications of this property.

In a sense, a quantum computer could be regarded as being a special kind of Multiple Instruction, Multiple Data (MIMD) parallel architecture. Indeed, the qubit can be represented as a vector of bits (multiple data) to which we can apply in parallel several unitary operators (multiple instructions).

In terms of quantum algorithm design, the key challenge is to exploit the intrinsic parallelism offered by a superposition of states.

1.1.7 QUANTUM COMPUTING PROPERTY #7

Quantum information cannot be copied. In other words, it is impossible to "copy" the superposition in one quantum register into another quantum register.

If we have a 2-qubit quantum register in an arbitrary superposition:

$$|\boldsymbol{R}\rangle = a|\boldsymbol{00}\rangle + b|\boldsymbol{01}\rangle + c|\boldsymbol{10}\rangle + d|\boldsymbol{11}\rangle \tag{1.67}$$

and Q is a 2-qubit quantum register in in a hypothetical Xerox machine, then there is no Xerox machine transformation U such that:

$$|\boldsymbol{R}\rangle|\boldsymbol{Q}\rangle \to U|\boldsymbol{R}\rangle|\boldsymbol{Q}\rangle = |\boldsymbol{R}\rangle|\boldsymbol{R}\rangle \tag{1.68}$$

for all possible quantum states $|\boldsymbol{R}\rangle$. This can be shown immediately by considering two arbitrary, normalized quantum states and assuming a unitary transformation that perfectly copies quantum states:

$$|\boldsymbol{\Psi}\rangle \otimes |s\rangle \quad \xrightarrow{U} \quad U\left(|\boldsymbol{\Psi}\rangle \otimes |s\rangle\right) = |\boldsymbol{\Psi}\rangle|\boldsymbol{\Psi}\rangle \tag{1.69}$$

$$|\boldsymbol{\Phi}\rangle \otimes |s\rangle \quad \xrightarrow{U} \quad U\left(|\boldsymbol{\Phi}\rangle \otimes |s\rangle\right) = |\boldsymbol{\Phi}\rangle|\boldsymbol{\Phi}\rangle\ . \tag{1.70}$$

Taking the inner products of the left-hand side terms of the two equations and the right-hand side terms of the two equations gives:

$$\langle\Phi||\boldsymbol{\Psi}\rangle = \langle\Phi||\boldsymbol{\Psi}\rangle^2 \tag{1.71}$$

which implies either

$$|\boldsymbol{\Psi}\rangle = |\boldsymbol{\Phi}\rangle \text{ or } |\boldsymbol{\Psi}\rangle \perp |\boldsymbol{\Phi}\rangle . \tag{1.72}$$

The above can only be guaranteed for states that are equal (inner product is unity) or are orthogonal (inner product is zero). The fact that it does not hold in general contradicts our assumption of a copy transformation U. This result is referred to as the *quantum no-cloning theorem.*

It is important to note that the no-cloning theorem forbids the production of *exact* copies of quantum states. Therefore, it is reasonable to suggest that a possible way around this limitation is to relax the requirement of perfect clones, and request instead *approximate* copies. Previous research has shown that it is possible to build an input-independent transformation, known as the Universal Quantum Copying Machine (UQCM), which is able to make approximate copies of arbitrary quantum states [60, 63]. It is possible to show that the following transformation is the most optimal UQCM for cloning a single qubit a into qubit b:

$$|\mathbf{0}\rangle_a|\mathbf{0}\rangle_b|\boldsymbol{Q}\rangle_x \rightarrow \sqrt{\frac{2}{3}}|\mathbf{00}\rangle|\uparrow\rangle + \sqrt{\frac{1}{3}}|+\rangle|\downarrow\rangle \tag{1.73}$$

$$|\mathbf{1}\rangle_a|\mathbf{0}\rangle_b|\boldsymbol{Q}\rangle_x \rightarrow \sqrt{\frac{2}{3}}|\mathbf{11}\rangle|\downarrow\rangle + \sqrt{\frac{1}{3}}|+\rangle|\uparrow\rangle \tag{1.74}$$

where:

$$|+\rangle = \frac{1}{\sqrt{2}}(|\mathbf{10}\rangle + |\mathbf{01}\rangle) \tag{1.75}$$

and $|Q\rangle_x$ is the quantum state of the copying-machine before the copying, while $|\uparrow\rangle$ and $|\downarrow\rangle$ are the states of the copying machine after the copying. Clearly, this transformation is only an approximate cloning device because the output also involves $|+\rangle$ states which were not present in the original input. Furthermore, the fidelity of the UQCM transformation can be characterized to be $2/3$. Indeed, if we perform a measurement to the output, we have a probability of $2/3$ to measure a and b in the same state.

Similarly, if the target qubit is in the general superposition given by:

$$|\boldsymbol{\Psi}\rangle = \alpha|\mathbf{0}\rangle + \beta|\mathbf{1}\rangle \tag{1.76}$$

then the use of a UQCM leads to:

$$|\boldsymbol{\Psi}\rangle_a|\mathbf{0}\rangle_b|\boldsymbol{Q}\rangle_x \rightarrow \alpha\left(\sqrt{\frac{2}{3}}|\mathbf{00}\rangle|\uparrow\rangle + \sqrt{\frac{1}{3}}|+\rangle|\downarrow\rangle\right) + \beta\left(\sqrt{\frac{2}{3}}|\mathbf{11}\rangle|\downarrow\rangle + \sqrt{\frac{1}{3}}|+\rangle|\uparrow\rangle\right) \tag{1.77}$$

This equation reveals that the approximate copies at the output are entangled among themselves and with the state of the copying machine. This is an unfortunate situation, as we cannot separate this state into the tensor product of three quantum states.

This approximate copying transformation can be generalized for the case of having K originals and M resulting clones for N-dimensional states [61, 62]. The fidelity of the output can be characterized through the following η parameter:

$$\eta = \frac{K}{M}\frac{M+N}{K+N} \tag{1.78}$$

Clearly, for a fixed K, as M and N grow large, the fidelity of the cloning machine rapidly decreases. Unfortunately, for most practical applications, we have a single original ($K = 1$), a substantial number of qubits n ($N = 2^n$), and we require several copies. Therefore, the use of UQCM often leads to low fidelity copies. Because of its several shortcomings, the applications of UQCM are very limited.

The inability to copy arbitrary quantum states is clearly limiting in terms of algorithmic flexibility. In particular, most nontrivial classical algorithms employ the use of copying into temporary variables to hold intermediate results.

However, the ability to copy classical states is permitted under the above analysis because the basis states are orthonormal. Indeed, remember that classical information is recovered when we use single state superpositions in the computational basis. These states are orthogonal and we can produce any number of exact copies out of it.

Another important consequence of Property #7 is that, if we have a quantum register in an unknown state, then we have no way to determine its value. Indeed, if we have:

$$|\boldsymbol{R}\rangle = \alpha|\mathbf{00}\rangle + \beta|\mathbf{01}\rangle + \gamma|\mathbf{10}\rangle + \delta|\mathbf{11}\rangle \tag{1.79}$$

and α, β, γ, and δ are unknown parameters, then upon measurement of the register we will obtain 00 with probability $|\alpha|^2$, and so on. If we did not have the no-cloning restriction, a trivial approach to determine these parameters would be to make several copies and measure all of them. Then, after applying some statistical methods we could estimate with an arbitrary degree of certainty the magnitude of these quantities. However, this is not possible in the quantum realm.

Because a quantum algorithm cannot copy quantum states, such operations must be avoided. An alternative to cloning is to initialize multiple quantum registers in the same state and applying the same operations to them until the point at which different operations need to be applied to different copies of the same state. However, this may be an inefficient, impractical, and expensive solution for most problems of interest.

1.1.8 QUANTUM COMPUTING PROPERTY #8

Quantum algorithms will always have to be initialized to the "0" position.

Any type of quantum algorithm will always start with a register initialized in the “0" position. The reason for this restriction is related to the side effect described before about how one cannot determine the state of a register if this is in an unknown state.

When the quantum computer is first booted, the quantum register may take an arbitrary quantum state which is impossible to determine. Thus, it is desired that, upon booting, the quantum register has to be placed on a known, pre-determined state. That is, the architecture of a quantum computer must be such that it guarantees that the initial state will always be the same.

This state is usually taken as the ground state of the underlying quantum system used to implement the quantum computer, and it is often used to represent "0". Therefore, in practice, the "0" state is achieved by allowing the quantum system that realizes the register to relax to its ground state. Of course, from a theoretical perspective the definition of "0" and "1" are completely arbitrary. However, because of engineering limitations, the ground state of the quantum register is defined to be "0", and it is required to be the only valid initial state of the computer.

This means that all quantum algorithms have to presume that "0" is the original state of the quantum register. However, by applying unitary transformations we can always obtain an arbitrary initial state. For example, if we require a uniform superposition of both values of a 1-qubit register, we apply a *Hadamard* transformation. The Hadamard gate is defined, in matrix form, as:

$$H = \frac{1}{\sqrt{2}} \begin{pmatrix} 1 & 1 \\ 1 & -1 \end{pmatrix} \tag{1.80}$$

which has the following effect on the 1-qubit $|\mathbf{0}\rangle$ state:

$$\begin{aligned} H|0\rangle &= \frac{1}{\sqrt{2}} \begin{pmatrix} 1 & 1 \\ 1 & -1 \end{pmatrix} \begin{pmatrix} 1 \\ 0 \end{pmatrix} && (1.81) \\ &= \frac{1}{\sqrt{2}} \begin{pmatrix} 1 \\ 1 \end{pmatrix} && (1.82) \\ &= \frac{1}{\sqrt{2}}(|0\rangle + |1\rangle) && (1.83) \end{aligned}$$

which is the desired uniform superposition of 1-qubit states.

For registers with more than 1-qubits we can apply one Hadamard gate for each qubit in the register in a tensor product form. For example, in the case of a 2-qubit register we have:

$$\begin{aligned} H^{\otimes 2}|\mathbf{00}\rangle &= H^{(1)} \otimes H^{(2)}|\mathbf{00}\rangle && (1.84) \\ &= H^{(1)}|0\rangle \otimes H^{(2)}|0\rangle && (1.85) \\ &= \frac{1}{\sqrt{2}}(|0\rangle + |1\rangle) \otimes \frac{1}{\sqrt{2}}(|0\rangle + |1\rangle) && (1.86) \\ &= \frac{1}{2}(|\mathbf{00}\rangle + |\mathbf{01}\rangle + |\mathbf{10}\rangle + |\mathbf{11}\rangle) && (1.87) \end{aligned}$$

which is a uniform superposition of all 4 states of the 2-qubit register.[4] In general, the matrix form for the Hadamard gate for n-qubit registers can be constructed using the convenient recursive formula:

$$H^{\otimes n} = \begin{pmatrix} H^{\otimes n-1} & H^{\otimes n-1} \\ H^{\otimes n-1} & -H^{\otimes n-1} \end{pmatrix}. \tag{1.88}$$

If instead of uniform superpositions we require more sophisticated initial states, we can still create them, but the initialization process may require a severe computational overhead. For instance,

[4]In some situations it is customary to use the symbol $|\mathbf{0}\rangle$ to represent the 0 state of a multi-qubit register. That is, instead of writing $|\mathbf{00...0}\rangle$ we just denote this ground state as $|\mathbf{0}\rangle$.

let us suppose that we have a n-qubit register that we want to initialize with a state $|\phi_{init}\rangle$, which is represented with 2^n complex parameters. Thus, we need to find a $2^n \times 2^n$ unitary matrix U that transforms the 0 state into the desired initial state. That is:

$$|\phi_{init}\rangle = U|\mathbf{0}\rangle \ . \tag{1.89}$$

To determine all the elements of the transformation matrix U, we basically need to solve a system of 2^n equations with 2^n variables. And to solve such a system of equations may require up to $O(2^n)$ computational steps.

Thus, even though it is theoretically possible to initialize a quantum register to an arbitrary state, this process may require an exorbitant amount of time. The initialization of a quantum register is, thus, a very important process that has to be accounted for when studying the computational complexity of quantum algorithms, and should never be ignored. Of course, if a problem requires the exact same initial state multiple times, the transformation matrix could be stored and be reused multiple times.

1.2 SUMMARY

We can summarize the properties of the quantum computational model as follows:

1. The qubit is the new unit of information - the classical bit is a special case.

2. Quantum computation is probabilistic - there is an intrinsic element of randomness involved when a measurement is made of a superposition of states unless the weight on one of the states is unity (i.e., all other weights are zero).

3. Measurement (read operations) are destructive. Even if a quantum superposition can index 2^n states, after measurement we only retrieve n-bits of logical classical information, and we cannot recuperate the original state.

4. The QC computational space is exponentially larger than the CC space for a fixed-size register - an n-qubit quantum register can simultaneously index 2^n n-bit states while its classical counterpart can only store the index of one n-bit state.

5. Transformations of quantum states must be reversible - an irreversible operation will cause the collapse of a superposition. Virtually, all classical logic gates can be efficiently replaced with functionally equivalent reversible gates.

6. QC offers intrinsic parallelism - the ability to simultaneously transform all states in a superposition permits some QC operations to be performed more efficiently than is possible for the best possible classical alternative.

7. Quantum information cannot be copied - the no-cloning theorem fundamentally constrains the class of algorithms that can effectively exploit quantum parallelism.

8. Quantum registers are always initialized to the 0 state.

CHAPTER 2

Advantages and Limitations of Quantum Computing

In the previous chapter we claimed that the classical computing (CC) framework is subsumed by the Quantum Computing (QC) model. This relationship can be made more explicit by recognizing that CC algorithms can be executed on a quantum computer as special cases in which all quantum registers store superpositions consisting of a single element, e.g.,

$$|\boldsymbol{R}\rangle = |\mathbf{00}\rangle \ . \tag{2.1}$$

In other words, if superpositions are never used, the quantum hardware simply implements a classical computer. Independently of whether quantum parallelism finds general use or is primarily applied to specialized problems, quantum computing hardware will eventually run all classical algorithms. This is because the effects of quantum phenomena increase as the size of logic gates decrease.

In this chapter we will discuss the specific advantages and disadvantages of the quantum computing model. We will also analyze a variety of algorithmic considerations when implementing a quantum algorithm on a simple quantum architecture.

2.1 QUANTUM COMPUTABILITY

In terms of algorithmic features, classical computing is subsumed by the quantum computing framework:

$$CC \subseteq QC \ . \tag{2.2}$$

That is, a quantum computer can do the same things classical computers can do, and perhaps much more. In other words, a quantum computer is at least as powerful as a classical computer.

On the other hand, it can be shown that a classical computer can (inefficiently) simulate a quantum computer. That is, if CC is augmented with a true random number generator, it is possible for any quantum algorithm to be simulated by a classical algorithm; however, the classical algorithm may require exponentially more computational resources. Thus, in terms of computability, whether a computational problem can be solved or not, classical and quantum computing are equivalent [49]:

- A function that can be computed using a classical computer can also be computed using a quantum computer.
- A function that cannot be computed using a classical computer cannot be computed with a quantum computer.

For example, it has been shown that the halting problem is not computable by a classical computer[1]; therefore, it is also incomputable using a quantum computer.

For the computer scientist, the principal distinction between QC and CC is that QC is more efficient than CC for specific classes of problems. Thus, the Church-Turing thesis remains valid:

> *Every function which would naturally be regarded as computable can be computed by a Turing Machine.*

However, the *strong* version of the thesis is contradicted by QC:

> *Every function which would naturally be regarded as computable can be computed* **efficiently** *by a Turing Machine.*

Specifically, there are problems that a Turing Machine cannot compute in time that is within a polynomial factor of what can be achieved using a quantum computer.

The violation of the strong Church-Turing Thesis means that new, meaningful, complexity classes can be defined for the quantum computing model.

2.2 CLASSICAL AND QUANTUM COMPLEXITY CLASSES

Let us recall that complexity classes are introduced in computer science to categorize how easy or how difficult is the solution of a problem. In particular, time and space complexity classes are defined independently of the underlying hardware used to solve the specific problem. As such, complexity classes are used to understand the difficulty of a set of problems, rather than benchmarking the performance of a hardware platform.

The most important classical and quantum computational classes are the following[2]:

- P = Class of problems that can be solved in polynomial time using a classical computer.
- NP = Class of problems whose solution can be verified by a classical computer in polynomial time.
- BPP = Class of problems that can be solved by a probabilistic algorithm on a classical computer in polynomial time with success probability of at least $1/3$.
- EQP = Class of problems that can be solved by a quantum computer in polynomial time with probability equal to 1.
- BQP = Class of problems that can be solved by a quantum computer in polynomial time with success probability of at least $1/3$.

[1] The halting problem can be stated as follows. Given a program that runs with a specific finite input, the problem is to determine if the program finishes the computation or runs forever.

[2] A more detailed description and analysis of classical complexity classes can be found in [16], while quantum complexity classes are discussed in [5].

It is important to note that the $1/3$ factor in these definitions is rather arbitrary. That is, we could have defined the class using any other number greater than 0 but smaller than 1. The reason is that the algorithm can be run a constant number of times to increase the probability of success. Then, the probability of error can be reduced to an arbitrarily small number.

As an example, the problem of computing the addition or any other arithmetic operation between two n-bit numbers is a problem that can be solved in $O(n)$ computational steps, and, therefore, it belongs to the complexity class P. On the other hand, *factorization* of a n-bit integer is a problem in NP; i.e., the most efficient known classical algorithm takes exponential time to compute a solution, but the solution can be verified in polynomial time. Within the realm of quantum computation, however, factorization can be computed in polynomial time (as we will further discuss in Chapters 5 and 6). Thus, factorization is a problem in both NP and BQP.

Although there are problems in both NP and BQP, there is no evidence that all problems in NP are in BQP. Interestingly, during the early days of quantum computing, it was hoped that NP was contained in BQP. However, the problems in NP $\cap$ BQP appear to have a special structure, so a sampled problem from NP is unlikely to be in BQP. At present the following hierarchies are known:

$$P \subseteq BPP \subseteq BQP \tag{2.3}$$
$$P \subseteq EQP \subseteq BQP\ . \tag{2.4}$$

Probably the two most important questions in complexity theory are:

$$NP \subseteq P\,? \tag{2.5}$$
$$NP \subseteq BQP\,? \tag{2.6}$$

In other words: Can all NP problems be solved efficiently by a classical computer? Can all NP problems be solved by a quantum computer? The consensus of researchers in complexity theory is that the answer is "no" to both questions.

At present what is known is that some problems can be solved more efficiently in QC than in CC, so the task of the computer scientist is to map out the classes of problems for which this is true. Clearly, the only way to achieve complexity reduction in QC is to exploit quantum parallelism, i.e., maintain and manipulate an exponentially large set of states in a quantum superposition (Properties # 4 and # 6).

2.3 ADVANTAGES AND DISADVANTAGES OF THE QUANTUM COMPUTATIONAL MODEL

To better understand the potential and limitations of quantum computing, consider the use of a quantum register to store a toy model of the entire universe. If the entire known universe, from the time of the Big Bang to date, is subdivided into a grid of Planck-scale size ($1.6 \times 10^{-35} m$) cells, each of which represents 1 bit of information, there would be a total of about 2^{800} bits of information. That amount information, therefore, could be stored in only 800 qubits!

Furthermore, an 800-qubit quantum register cannot only hold a discretized representation of the entire universe, it can be transformed in $O(1)$ time. In other words, a transformation U can be applied to all 2^{800} grid cells simultaneously.

Does this mean that a quantum computer can actually simulate the universe? The answer is *no*, and the reasons illustrate the limitations of the QC framework:

1. *Initialization* - Our toy model represents the universe as a grid in which the cell-size is at the Planck limit of information. To define the state requires an assignment of 0 or 1 values to each of the 2^{800} cells. This can be done by performing 2^{800} explicit assignment operations to the quantum register, but doing so would take trillions and trillions... and trillions of years. Alternatively, a function $f(x)$ can be applied to simultaneously assigned values to all cells indexed by the parameter x in $O(1)$ time (assuming that the evaluation of $f(x)$ takes only $O(1)$ time per cell).

2. *Transformation* - In the same way that an arbitrary function $f(x)$ can be applied to initialize all the cells in parallel, a function can also be applied to transform all the cells in parallel. It would seem, then, that some kind of Markov model could be applied to simulate the evolution of the state of the universe. Unfortunately, within QC it is impossible to apply a function to a particular state which depends on other states in a superposition. For example, a function of the type $f(x, y)$, where x and y are two different computational cells. In other words, each state must be either transformed independently or other states must be stored for access outside of the superposition. In the former case we are limited to rather uninteresting models, and in the latter case we must store $O(2^{800})$ classical bits of information, which is clearly impractical.

3. *Output* - Even if we can provide useful functions to initialize and evolve our model of the universe, we cannot extract any information from the simulation until we read the state of the quantum register. At that point the superposition collapses and the register stores the state of one of the 2^{800} cells. In other words, we may simulate the evolution of 2^{800} states but can only access one of them. That is, we can only output 800 bits of logical information. It is possible to re-run the simulation multiple times to sample more states, but there is no way to avoid the fact that meaningful statistics will require $2^{800}/O(1)$ computations. As will be described in a subsequent chapter, optimal QC computation of the mean of N states in a superposition requires $\Theta(N^{1/2})$ time, which translates to a prohibitive number of computations proportional to 2^{400} for our toy model of the universe.

This simple, and perhaps unrealistic, example illustrates how the main advantage of the quantum computational model is the intrinsic computational parallelism over an exponentially large computational space (Properties # 4 and # 6). And the principal disadvantages of the quantum model are the inability to make copies and the destructive nature of quantum measurement (Properties # 3 and # 7).

To further appreciate the constraints imposed by the model, let us consider the pseudocode for the following algorithm:

$$
\begin{aligned}
1: &\quad print\ a; \\
2: &\quad c = b; \\
3: &\quad f(a); \\
4: &\quad f(b); \\
5: &\quad f(c);
\end{aligned}
$$

This algorithm cannot be translated into a quantum version if a, b, and c are quantum variables intended to store arbitrary superpositions of states. In particular, the *print* statement can be interpreted as a read or measurement operation which collapses the quantum superposition in a, and thus $f(a)$ becomes meaningless. Also, we cannot copy the value of the quantum superposition b onto c in such a way that both $f(b)$ and $f(c)$ make sense. Of course, depending on the underlying quantum architecture and the semantics and syntax of the quantum language, one could interpret line 2 of the pseudo-code as a reassignment of addresses rather than an actual copy. But even so, lines 4 and 5 imply different computational paths for b and c. In any event, this pseudo-code makes evident the algorithmic restrictions of the quantum model.

These limitations severely restrict the class of algorithms for which quantum computing can provide better algorithmic performance than the classical model. Thus, on the one hand it appears that quantum computing is so powerful that a single quantum computer can represent the state of the entire universe. On the other hand, it appears that nothing useful can be done with that representation.

Of course, we could propose a trivial way to avoid these issues. By simply avoiding the use of quantum superpositions of more than one state we could overcome the no-cloning restriction, as well as, the destructive nature of quantum measurements. But in this case, we would be working with the classical model of computation. That is, we would be running classical algorithms on quantum hardware. If our quantum computer supports the use of large quantum superpositions, then we would be wasting computational resources.

Therefore, the big question is: what differences between a quantum computer and a classical computer can be verified? Secondly, how can these differences be exploited to produce more efficient solutions to particular problems?

Superficially, it seems that the rules of quantum computing, e.g., destructive measurements and no-cloning, conspire to make it impossible to prove whether or not a superposition provides any computational advantage. If this were truly the case, quantum parallelism would have no practical value. As will be shown, however, the quantum model offers optimal solutions for certain classes of problems.

Finally, it is important to observe that these algorithmic disadvantages of the quantum model are actually seen as huge advantages in the context of cryptography and secure communications. Indeed, the no-cloning theorem means that information cannot be forged, and the destructive nature of quantum measurement implies that it is very easy to determine if an eavesdropper is trying to intercept a secure communication [40, 7, 13].

2.4 HYBRID COMPUTING

As has been discussed, in the long term quantum circuits will comprise all computing hardware. The more significant question is whether quantum parallelism will play a major or minor role in algorithm design. It is possible that quantum parallelism provides complexity advantages only for a narrow range of practical problems. On the other hand, it may provide speedups for a much wider range of algorithms when used as an auxiliary tool – much like vector processors are with some classical computers.

We use the term "hybrid computing" to refer to the process of developing algorithms by treating a quantum register as a special-purpose resource which permits certain operations to be performed efficiently [32]. This approach is somewhat less general than opening up the entirety of quantum logic to the design of new algorithms. On the other hand, it also dramatically reduces the difficulty of deriving new algorithms by focusing on the exploitation of known efficient solutions for generic classes of problems. This is what is done in classical algorithm design when, for example, a solution to a problem is tailored so that binary search can be applied.

The motivation for establishing the hybrid computing model is that it minimizes the intrusion of the details of quantum phenomena into the algorithm design process. This allows the computer scientist to exploit the power of quantum computing in the same way that modules in a classical software library are used. The distinction between hybrid algorithm design and full quantum algorithm design can be summarized as follows:

> Hybrid algorithm design consists of identifying the best possible algorithmic solution built from a set of given classical and quantum building blocks. Full quantum algorithm design, by contrast, also permits the development of new building blocks that cannot be defined in terms of other building blocks/primitives.

In addition to simplifying the process of algorithm design, the hybrid computing model also suggests a simplified notional computer architecture.

2.5 THE QRAM ARCHITECTURE

The Quantum Random Access Machine (QRAM) hybrid architecture conveys the fact that the quantum processor is used only to speed up certain subroutines, functions, or computational kernels [55]. In the QRAM architecture, shown in Figure 2.1, the classical and quantum processors work in a master/slave fashion. Here, classical code makes calls to an external device, the quantum processor. In addition, quantum code written in some adequate quantum language provides the instructions to be carried out by the quantum processor. Thus, the classical computer loads the quantum code, and sends a stream of instructions and data to the quantum processors.

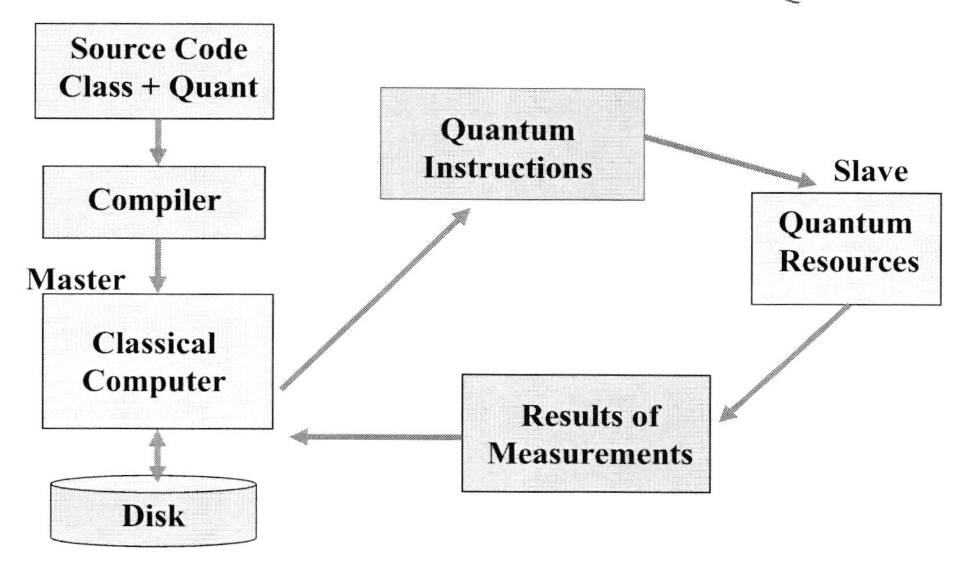

Figure 2.1: The QRAM Architecture.

Once these instructions sent by the classical computer have been carried out in the quantum processor, a measurement is performed and the result is sent back to the classical processor. And this process may be repeated as a pipelined cycle multiple times.

Clearly, in this architecture the quantum processor incurs the same limitations as more familiar augmentations, such as, vector and graphics processors, e.g., data transfer overhead. In addition to these limitations, the quantum processor potentially introduces others. Depending on how the quantum register is implemented, there may be a limit on the amount of time that a quantum superposition can be maintained. This *decoherence* time represents a constraint which has no parallel in classical algorithm design[3].

2.5.1 ALGORITHMIC CONSIDERATIONS

As we have discussed before, a quantum algorithm that exclusively uses nontrivial quantum superpositions has many potential problems because of the no-cloning and destructive measurement restrictions. And on the other hand, a quantum algorithm that only uses trivial superpositions does not take advantage of the algorithmic structure of quantum computing. Clearly, there is an issue of algorithmic balance, and the quantum software engineer will have to determine which portions of his code are good candidates for quantum acceleration.

The design of effective quantum algorithms will not be easy, as there are several algorithmic and architectural considerations that need to be taken into consideration. For instance, it still remains to be

[3]However, there is a parallel to the constraints imposed by signal propagation delays when implementing classical computing hardware.

determined the expected theoretical speed of the first generations of quantum processors. However, it is very likely that these quantum processors will be much slower than their classical counterparts. A good estimate is that quantum processor speeds will be in the order of MegaHertz, compared to a predicted speed on the order of hundreds of GigaHertz expected on classical computers by the year 2025. Therefore, the quantum software engineer has to be sure that the slow quantum hardware does not override the algorithmic improvements of the quantum model.

Also, as already discussed, the quantum processor may be subject to short decoherence times. From an algorithmic point of view this means that the software engineer will have to estimate the running time of the program and be sure it will be shorter than the typical decoherence time of the quantum processor he is working with. This may mean that for some long computations, the problem may have to be broken in several parts to guarantee that decoherence will not interrupt the computation.

Also, the QRAM architecture described above implies the transference of data and instructions from the classical processor to the quantum block and the transmission of raw data obtained from quantum measurements on the reverse route. This processes will clearly involve some data transfer overheads. Thus, if the datasets are large, this data transfer step may take a long time, and the advantage of quantum acceleration could completely disappear.

The Input/Output (I/O) characteristics of the quantum computational model also bring algorithmic considerations that need to be accounted for. From Property #8, a quantum computer can only be initialized in the 0 state. Of course, we can always use gates to transform the 0 state into any initial state desired by the software designer. However, if this initial state is arbitrary, there may be need to perform $O(poly(2^n))$ computational steps to find the reversible unitary gate that makes the desired transformation. Once more, the computational advantage of the quantum model may be overridden by this factor.

For most quantum algorithms, the initial state is required to be an uniform superposition. As we discussed on the previous chapter, this can be accomplished by applying n Hadamard gates. Thus, these algorithms carry an overhead of $O(n)$ for the initialization process. However, this overhead is likely to depend on the hardware architecture of the quantum computer, as one could easily imagine the implementation of multi-dimensional Hadamard gates. For such an architecture, the initialization overhead would effectively be reduced to $O(1)$.

Regarding output, let us recall that because of the destructive nature of quantum measurement, we cannot output the entire superposition. As the reader will remember, we can get a specific state with some probability. Thus, a n-qubit register that holds 2^n states can only output n-bits of logical information. Depending on the algorithm, some information may get lost in the process. If there is a need to extract more than n-bits of classical information, the user could perform the exact same algorithm repeated times. However, once more, this process could easily overwhelm the computational advantage granted by quantum acceleration.

Memory addressing schemes may also turn to be problematic. Let us consider a n-qubit quantum register. The most trivial memory-addressing scheme that comes to mind, in the form of a binary tree traversing the states of the quantum register, requires $O(n)$ computational steps to dereference an address and uses $O(2^n)$ switching gates [40]. Unfortunately, this model is highly inefficient and consumes a large number of computational resources. However, faster implementations have been proposed [50].

At the same time, it is unrealistic to expect that quantum computing will work efficiently with existing, pre-computed data. Indeed, if we have a n-qubit register that points at 2^n states and we need to populate it with some data that is stored in disk space somewhere else, then we require $O(2^n)$ operations to move the data from disk to memory. For a large system, of say, 800 qubits, such an operation becomes completely unfeasible.

Therefore, in general it appears that quantum computing is better suited to work with dynamically generated data sets. In this case, there is no need to move precomputed data from two locations within the computer. However, the problem, in this case, goes back to the impossibility to export the entire data set.

Therefore, it will be a truly challenging task for quantum software engineers to determine which portions of a classical program are good candidates for quantum acceleration. Most probably, they will have to rewrite the entire code from scratch. As such, the optimization of hybrid programs is a big challenge that should be addressed before the deployment of quantum computers. It is interesting to note, however, that a strikingly similar problem is found in reconfigurable supercomputing using FPGAs[4].

2.5.2 QUANTUM ALGORITHM DESIGN

At this point we can enumerate the most basic issues that need to be considered when designing quantum algorithms:

1. To design a quantum algorithm we need to take advantage of the parallelism and exponentially large computational space of QC.

2. We also need to avoid instances where we need to copy or read data out of the register.

3. We need to be very careful about probabilistic and reversibility issues.

4. We have to keep in mind the algorithmic considerations that may arise from a specific QC hardware architecture, such as, the QRAM model.

5. Finally, we need to provide optimality results that prove a better quantum performance than with the best known classical method.

[4]A Field Programmable Gate Array (FPGAs) is a digital integrated circuit that can be configured and programmed by the user to perform a variety of computational tasks. The hardware configuration of this circuit takes place after the manufacturing process, which means that the user is no longer limited to a fixed, unchangeable, and predetermined set of hardware functions. In principle, any Boolean function can be mapped into an FPGA, which offers the possibility of increasing the computational performance of a code by means of pipelining, parallelism, and concurrency [56].

By any means, the design of effective quantum algorithms is not an easy task. Because of such intricacies, perhaps it should not be a surprise that there are only a handful of efficient quantum algorithms reported in the open literature.

2.6 QUANTUM BUILDING BLOCKS

At present there are six general algorithmic areas for which the quantum model appears to offer substantial benefits when compared to the best known classical alternatives:

1. *Amplitude Amplification* - This procedure permits the weight associated with a desired state within a quantum superposition to be increased (amplified) so that it is more likely to be measured. Amplitude amplification can be used as a generic tool for efficiently finding a solution state for a wide variety of search and optimization problems.
2. *Quantum Fourier Transform (QFT)* - The QFT permits certain restricted types of Fourier transformation information to be computed in time sublinear with the size of the dataset. The best classical algorithm require superlinear time.
3. *Quantum Random Walks* - QRW can be applied to efficiently solve a wide variety of statistical estimate problems which cannot be efficiently simulated with a classical algorithm. A very detailed introduction to use of quantum random walks in computer science can be found in [52].
4. *Quantum Error Correction* - QEC is analogous to classical error correcting techniques except that it can recognize and correct qubit errors, as opposed to classical bit flips [40]. Its applications are in the implementation of the QRAM architecture rather than in the hybrid algorithm design process.
5. *Quantum Cryptography* - Quantum cryptography is a major area of research, but it finds few if any applications within the QRAM computing architecture [40, 7].
6. *Simulation of Physical Systems* - It should not be surprising that quantum phenomena can be simulated more efficiently on quantum hardware than is possible classically. In fact, this application was what first motivated the study of "quantum computing" by Richard Feynman [17]. In recent years, there has been a surge of adiabatic quantum algorithms to solve challenging problems in bioinformatics, such as, protein docking and protein folding [4, 25]. While this application of quantum computers promises revolutionary advances in a variety of scientific and engineering areas, it is not very interesting from a formal perspective. Indeed, mapping the evolution of a physical system to the evolution of the quantum computer is somewhat equivalent to using a water tank to simulate ocean dynamics.

For the computer scientist, the first three of the above are the most relevant for algorithm design. Amplitude Amplification and QFT will be described in more detail in the following chapters, while QRW are well described in another volume of the series [52].

2.7 SUMMARY

By far, the greatest advantage of the quantum computational model is the ability of performing operations in parallel to an exponentially large superposition. Unfortunately, the model is restricted by the destructive nature of measurement operations and by the impossibility of making exact copies of quantum information. Interestingly, while these last two features are detrimental to the algorithmic capabilities of quantum computing, these are powerful advantages for the application of quantum information to secure communications. Indeed, the no-cloning theorem implies that quantum information cannot be forged, while the destructive nature of measurements implies that we can easily detect eavesdroppers.

Furthermore, the performance of a quantum algorithm will also be limited by a number of factors related to the specific architecture used to implement the quantum computer. In this regard, quantum algorithm development should consider the overheads involved in I/O operations, memory retrieval, and initialization.

To date, the most important building blocks to develop quantum algorithms are amplitude amplification techniques, the Quantum Fourier Transform, and quantum random walks.

CHAPTER 3

Amplitude Amplification

In the previous chapter we discussed the way in which quantum physics seemingly conspires against any attempt to exploit the supposed storage of an exponentially large number of states in a quantum superposition. Specifically, any extraction of information from a quantum register causes the superposition to collapse to a single state, at which point it is equivalent to a classical state in a classical register.

In this chapter we will demonstrate that the notion of a superposition of states can be exploited to obtain solutions to many classes of search and optimization problems more efficiently than is possible with the best available classical alternatives. This is achieved using a technique referred to as *amplitude amplification*, and it basically consists of transforming a superposition so that desired states are more likely to be measured than are other states [10].

Without loss of generality, we will assume that we have a Boolean function, which takes a bit vector as an input and then outputs a 0 or a 1 depending on whether the input vector represents a solution. More technically, we will be considering the case of a given Boolean indicator function:

$$\chi : \{0, 1\}^n \to \{0, 1\} \tag{3.1}$$

which partitions the set

$$X \equiv \{0, 1\}^n \tag{3.2}$$

into the set of solutions and the set of nonsolutions where

$$\chi(x) = \begin{cases} 0 & x\text{-is-not-a-solution} \\ 1 & x\text{-is-a-solution} . \end{cases} \tag{3.3}$$

This function is referred to as an *oracle*, i.e., a black box that can tell if x is a solution or not in constant time $O(1)$. Let us consider now a quantum algorithm A [1]:

$$A|\mathbf{0}\rangle = \sum_{i=0}^{2^n-1} \alpha_i |\boldsymbol{i}\rangle . \tag{3.4}$$

Now, the application of the oracle χ to this state will basically assign to each $|\boldsymbol{i}\rangle$ a value 0 or 1 (depending on whether or not it is a solution) and this state will have probability α_i^2 of being measured. Therefore, if the sum of the probabilities associated with solution states is a:

$$a = \sum_{i=0}^{2^n-1} |\alpha_i|^2 \chi(i) \tag{3.5}$$

[1] Here, we presume that A is the concatenation of all possible unitary operations involved in the algorithm.

then a measurement applied to the superposition will give a solution with probability a, which means the expected time to find a solution is $O(1/a)$. Indeed, if the probability of measuring a solution is $1/3$, then we expect to perform the same procedure about 3 times to extract a solution.

Measuring the above superposition of states is essentially a random search algorithm, which will require expected $O(1/a)$ repetitions of the procedure. Let us consider the case where A generates a uniform superposition. Then, there are N states in a uniform superposition of n-qubits. If there is only one solution state, then the probability of measuring it is $1/N$, and the expected complexity is, therefore, $O(N)$. This complexity is no better than a classical exhaustive examination of all the states. The purpose of amplitude amplification is to transform the superposition of states so that $a \approx 1$, i.e., the probability of measuring the solution state is much higher than $1/N$.

3.1 QUANTUM SEARCH

General and provably optimal algorithms exist for performing amplitude amplification to identify a desired solution from a superposition of states. Specifically, it applies $O\left(N^{1/2}\right)$ transformations to the superposition of states in a quantum register so that a subsequent measurement will obtain the desired solution state with high probability. In other words, quantum search provides a quadratic speedup over classical exhaustive search.

3.1.1 QUANTUM ORACLES

A quantum oracle is a quantum implementation of the oracle described on the previous section. Clearly, any classical functional criteria can be described by means of a classical oracle. And any classical oracle can be implemented in the quantum domain by simply adding extra qubits to guarantee the reversibility of the operation.

In such a case, the effect of a quantum oracle O for a solution $f(x) = 1$ for an indicator Boolean function f is:

$$|\boldsymbol{x}\rangle|\boldsymbol{q}\rangle \xrightarrow{O} |\boldsymbol{x}\rangle|\boldsymbol{q} \oplus \boldsymbol{f}(\boldsymbol{x})\rangle \tag{3.6}$$

where $\oplus$ indicates module 2 addition. By choosing the right input state, we can conveniently rewrite this expression as:

$$|\boldsymbol{x}\rangle\left(\frac{|\boldsymbol{0}\rangle - |\boldsymbol{1}\rangle}{\sqrt{2}}\right) \xrightarrow{O} (-1)^{f(x)}|\boldsymbol{x}\rangle\left(\frac{|\boldsymbol{0}\rangle - |\boldsymbol{1}\rangle}{\sqrt{2}}\right) . \tag{3.7}$$

Clearly, the effect of the oracle is to change the phase of the solution, while leaving unchanged the rest of the elements of the superposition.

In the above formula we have added an extra qubit to the register in the form of $|\boldsymbol{q}\rangle$. This has actually helped in moving the information of the quantum oracle into a phase change. However, as the state of the extra qubits remain unchanged after the application of the oracle, these can be safely ignored. Then, the effect of the oracle is often written as:

$$|\boldsymbol{x}\rangle \xrightarrow{O} (-1)^{f(x)}|\boldsymbol{x}\rangle . \tag{3.8}$$

Therefore, the matrix representation of the oracle operator O has -1 in the diagonal term for the solution, 1 for the other diagonals, and zeros in the off-diagonals. For two qubits ($N = 4$), the matrix for the oracle where 00 is the solution to the search problem is:

$$\mathrm{O}_{00} = \begin{pmatrix} -1 & 0 & 0 & 0 \\ 0 & 1 & 0 & 0 \\ 0 & 0 & 1 & 0 \\ 0 & 0 & 0 & 1 \end{pmatrix} . \tag{3.9}$$

It is important to remark that we can write this matrix because for simulation purposes we have assumed that the solution is 00. However, in a real case scenario the solution to the search problem is obviously an unknown quantity. Thus, this type of matrix representation for the oracle is only useful when we try to simulate or represent the effects of the oracle. When a real quantum computer implements an oracle, only the oracle knows the exact form of this matrix, and we do not have direct access to it. In other words, the determination of the matrix representation of the oracle is equivalent to solving the search problem.

It is also important to remark that, in spite of its mystical name that invokes scenes of magical creatures from the ancient Greek lore, there is nothing supernatural or mysterious about the concept of an oracle. Clearly, an oracle can always be used in classical computing to solve a search problem, and the application of an oracle is ultimately equivalent to the computation of the function f for a given value of the input.

The same idea remains valid in the quantum domain. The difference is, of course, that a quantum oracle can be applied in parallel to all the possible values of the input described by the quantum register. That is, if we have an arbitrary superposition:

$$|\boldsymbol{\Psi}\rangle = \sum_{i=0}^{2^n-1} \alpha_i \, |\boldsymbol{i}\rangle \tag{3.10}$$

then the application of an oracle O results in:

$$O|\boldsymbol{\Psi}\rangle = \sum_{i=0}^{2^n-1} (-1)^{f(i)} \alpha_i \, |\boldsymbol{i}\rangle \tag{3.11}$$

in a single computational step. In other words, in the quantum computing model we can perform 2^n simultaneous checks to a n-bit oracle.

In addition, the requirement that the oracle is a black box function does not mean that the implementation is difficult or unknown. It only means that we can assume that the computational complexity of computing f for some input value x is taken to be $O(1)$. This way, whatever possible computational complexity involved with the computation of f is isolated from the computational complexity of the search algorithm itself. In this case, the overall computational complexity of the algorithm, reveals the complexity of the general search problem, independently of the sophistication of the particular search criteria.

3.1.2 SEARCHING DATA IN A QUANTUM REGISTER

An n-qubit quantum register can hold a quantum superposition of $N = 2^n$ states. The states are just the set of possible n-bit binary vectors, so we can interpret them as the integers from 0 to $N - 1$. Now suppose that we are interested in knowing whether any of these numbers has some special property assessed by our oracle, i.e., $f(x) = 1$ says that x has the property. Clearly, a classical computer requires $O(N)$ computational checks to find this element. More specifically, a classical computer requires to use the oracle $O(N)$ times.

In generally, a quantum oracle can be applied to an array of N data items, and the n-bit binary vectors in the superposition can be interpreted as indices into the array. For example, the zero-vector state would refer to the first element of the array. With this we can now perform a quantum search for an element of the array that has a desired property. In other words, we can perform a general quantum search of any kind of dataset.

Before we present the mechanics of quantum searching, it is useful to consider what a quantum search is expected to do. We are hoping to exploit quantum parallelism, which means that we want to simultaneously examine all of the states in our superposition to find what we are looking for.

However, it seems that our dataset array represents a bottleneck because our oracle can only access one element at time. Indeed, at first sight it appears that we would need N copies of the oracle and array to permit it to be applied simultaneously to each of the N states. Amazingly, the intrinsic parallelism of the quantum computing model enables a superposition of the oracle for every possible state of the register. In other words, N indented executions of the oracle for each of the N states in the quantum register will occur simultaneously.

Quantum parallelism in this case appears to do the impossible. The application of the oracle to the quantum register appears to be just one operation, so where and when do all $N = 2^n$ oracle executions take place? The answer is that they all happen when the oracle is applied to the quantum register. There is little else that can be said. It is this kind of strong deviation from ordinary experience that makes the prospects for quantum computing so interesting.

Before concluding that quantum physics offers power comparable to magic, it is important to keep in mind that we cannot obtain any information from the processing of a quantum register until we make a measurement, and at that instant the superposition collapses. We can apply our oracle to test all N data items simultaneously. As we just saw, the oracle will shift the phase of the solution leaving unchanged the rest of the elements.

For example, if we have an uniform superposition of all $N = 2^n$ elements of a n-bit quantum register:

$$|\Psi\rangle = \frac{1}{\sqrt{N}} \sum_{i=0}^{N-1} |i\rangle \tag{3.12}$$

and we apply the oracle to get:

$$O|\Psi\rangle = \frac{1}{\sqrt{N}} \sum_{i=0}^{N-1} (-1)^{f(i)} |i\rangle . \tag{3.13}$$

At this point the solution is marked by having a phase shift; however, all elements are equally likely to be read. Thus, the question is, how do we determine which data item represents a solution?

To be able to identify solution states, we could augment our quantum register with an extra qubit which will mark a state as 1 or 0, depending on whether the oracle identifies it as a solution or not. Thus, if we apply the oracle to the superposition of states and then take a measurement, we will know whether or not the measured state is a solution. Unfortunately, this does not allow us to find the solution any more efficiently, because all we can do is repeat the whole process until we measure a state that is marked as being a solution. Doing this takes $O(N)$ iterations, which is no more efficient than classical search.

At this point it appears that quantum computing offers an algorithm that is no better than a classical algorithm, which repeatedly chooses a random element of the array and tests it with the oracle. In other words, it appears that the potential power of quantum parallelism evaporates under closer examination. This is not the case, however, because a more sophisticated algorithm can be applied to manipulate the superposition so that a solution state is the most likely to be measured. This algorithm is referred to as *Grover's algorithm* [23, 24].

3.1.3 GROVER'S ALGORITHM

To illustrate Grover's algorithm, we will consider the case of searching for the index of a desired data item y from a dataset of size $N = 2^n$ data items. We presume that the dataset is completely unordered and unstructured.

The first step is to create a uniform superposition of the N indices in the quantum register, where *uniform* simply means that the weights associated with the states are equal (and, of course, sum to unity). As discussed on a previous chapter, this can be done using Hadamard gates. Thus, we initialize the n-qubit register to the uniform superposition as:

$$|\boldsymbol{\Psi}\rangle = H^{\otimes n}|\mathbf{0}\rangle = \frac{1}{\sqrt{N}} \sum_{i=0}^{N-1} |\boldsymbol{i}\rangle \tag{3.14}$$

where each state in the superposition is an index into the array of data items to be searched.

As before, we can observe that at this point the solution is already in the quantum register, but direct measurement will lead to the solution with a very small probability of success. Indeed, if we measure the quantum state $|\boldsymbol{\Psi}\rangle$, the probability that the result will be the unique solution is $1/N$. This can be seen from the following:

$$\xi = \langle \boldsymbol{y}|\boldsymbol{\Psi}\rangle = \frac{1}{\sqrt{N}} \sum_{i=0}^{N-1} \langle \boldsymbol{y}|\boldsymbol{i}\rangle = \frac{1}{\sqrt{N}} \tag{3.15}$$

where the probability of success is $\xi^2 = 1/N$. The goal is to modify $|\boldsymbol{\Psi}\rangle$ to amplify the amplitude ξ, and the probability of success ξ^2, for the state $\boldsymbol{i} = y$.

The necessary modification of the quantum state involves the use of the oracle O and another quantum operator $\boldsymbol{D}$ that inverts states through the mean. The diagonal elements of $\boldsymbol{D}$ are $-1 + 2/N$, and the off-diagonals are $2/N$. The Grover iteration operator is therefore defined as:

$$G = D \times O \,. \tag{3.16}$$

For example, in the case of two qubits ($N = 4$), D is:

$$D = -\left(\frac{1}{2}\right)\begin{pmatrix} 1 & -1 & -1 & -1 \\ -1 & 1 & -1 & -1 \\ -1 & -1 & 1 & -1 \\ -1 & -1 & -1 & 1 \end{pmatrix} \tag{3.17}$$

and if the solution state in this example is 00, then the oracle looks like:

$$O_{00} = \begin{pmatrix} -1 & 0 & 0 & 0 \\ 0 & 1 & 0 & 0 \\ 0 & 0 & 1 & 0 \\ 0 & 0 & 0 & 1 \end{pmatrix} . \tag{3.18}$$

In this case, the Grover operator G results:

$$G = D \times O_{00} = -\left(\frac{1}{2}\right)\begin{pmatrix} 1 & 1 & 1 & 1 \\ -1 & -1 & 1 & 1 \\ -1 & 1 & -1 & 1 \\ -1 & 1 & 1 & -1 \end{pmatrix} . \tag{3.19}$$

The initial uniform superposition for this simple example can be written as:

$$|\mathbf{\Psi}\rangle = \frac{1}{2}\left(|\mathbf{00}\rangle + |\mathbf{01}\rangle + |\mathbf{10}\rangle + |\mathbf{11}\rangle\right) = \frac{1}{2}\begin{pmatrix} 1 \\ 1 \\ 1 \\ 1 \end{pmatrix} . \tag{3.20}$$

The amplitude and probability for each element in the superposition can be visualized in Figure 3.1. As made evident by this graphical representation, all amplitudes are the same, and as a consequence, all the probabilities are equal.

After the application of the oracle operator to the uniform superposition, the state of the quantum register looks like:

$$O_{00}|\mathbf{\Psi}\rangle = \begin{pmatrix} -1 & 0 & 0 & 0 \\ 0 & 1 & 0 & 0 \\ 0 & 0 & 1 & 0 \\ 0 & 0 & 0 & 1 \end{pmatrix}\begin{pmatrix} 1 \\ 1 \\ 1 \\ 1 \end{pmatrix}\frac{1}{2} = \frac{1}{2}\begin{pmatrix} -1 \\ 1 \\ 1 \\ 1 \end{pmatrix} . \tag{3.21}$$

As expected, only the phase of the solution state has been changed by the application of the oracle. Therefore, the phase of the solution is inverted, but the probabilities remain the same. This is visualized in Figure 3.2

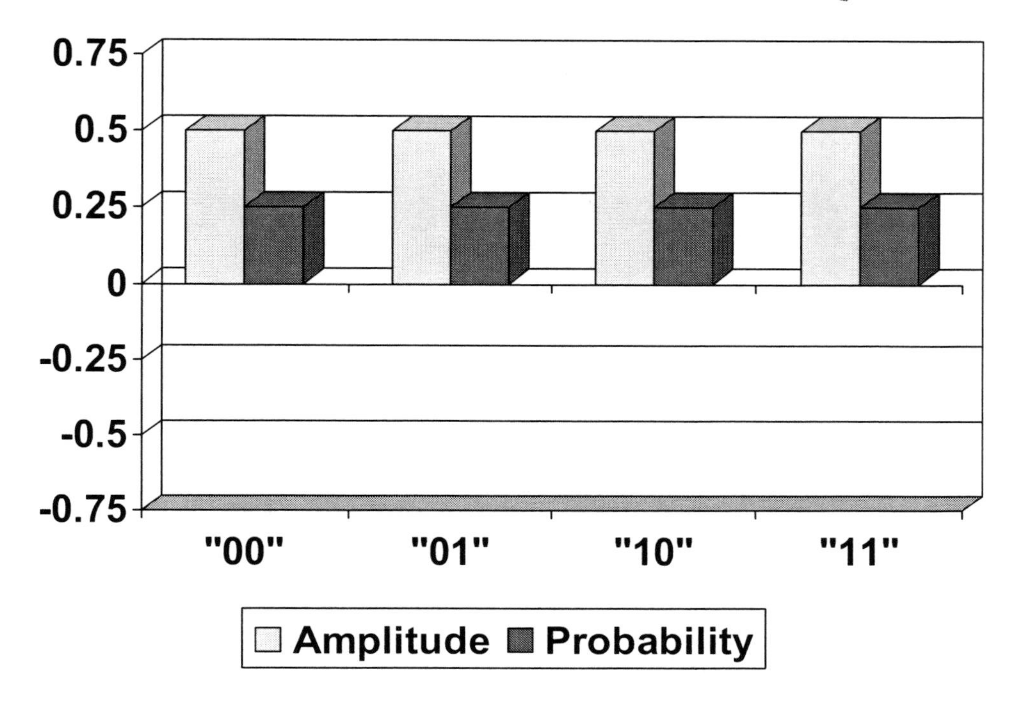

Figure 3.1: Visualization of the initial state in Grover's Algorithm for 2 qubits, and "00" as the solution to the search problem. At this stage, all the elements of the superposition have the same amplitude and probability.

The next step is to apply the inversion around the mean operator D. First let us notice from Figure 3.2 that the mean of the four amplitudes is:

$$mean = \frac{-\frac{1}{2} + \frac{1}{2} + \frac{1}{2} + \frac{1}{2}}{4} = \frac{1}{4} \,. \tag{3.22}$$

Thus, for "00" we have an amplitude of -0.5, and its "distance" from the mean is 0.75. Then, an inversion around the mean implies to invert -0.5 around 0.25, which leads to $0.25 + 0.75 = 1.0$. On the other hand, for the states "01", "10", and "11", they all have an amplitude of 0.5, and their distance from the mean is 0.25. Then, the inversion around the mean for these states implies to invert 0.5 around 0.25, which leads to $0.25 - 0.25 = 0.0$ for all of them. The effect of the inversion around the mean is depicted in Figure 3.3.

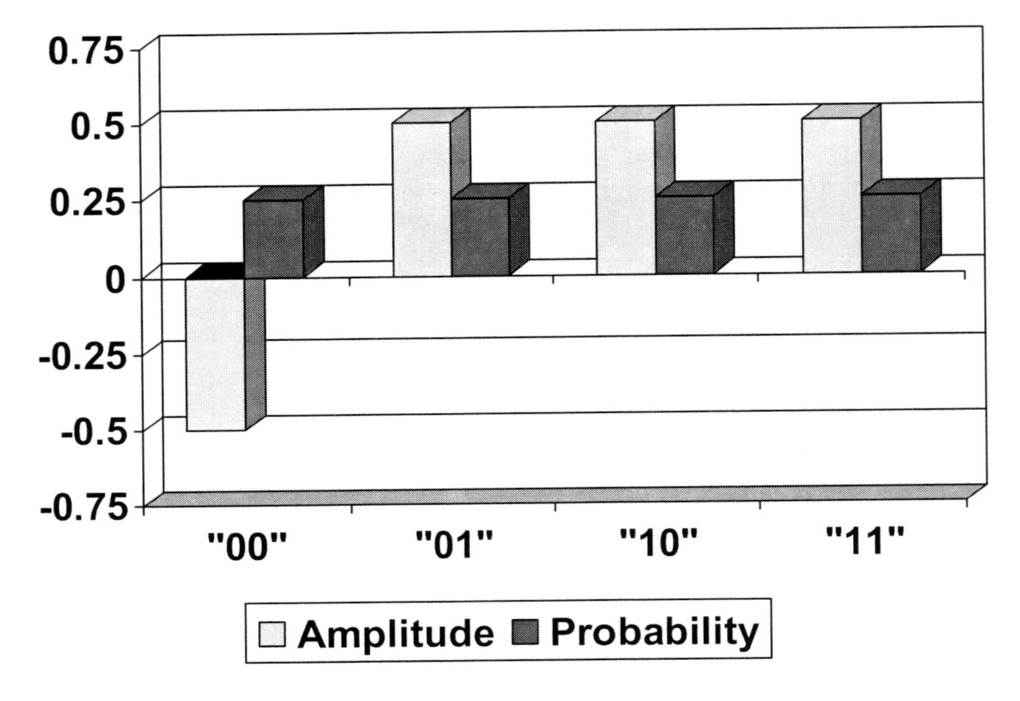

Figure 3.2: Visualization of the application of the oracle in Grover's Algorithm for 2 qubits, and "00" as the solution to the search problem. At this point, the amplitude for the state '00" has shifted its phase and acquired a minus sign. The probabilities remain the same for all the states.

Equivalently, we could have used the operator D:

$$G|\Psi\rangle = DO_{00}|\Psi\rangle \tag{3.23}$$

$$= -\left(\frac{1}{4}\right)\begin{pmatrix} 1 & -1 & -1 & -1 \\ -1 & 1 & -1 & -1 \\ -1 & -1 & 1 & -1 \\ -1 & -1 & -1 & 1 \end{pmatrix}\begin{pmatrix} -1 \\ 1 \\ 1 \\ 1 \end{pmatrix} \tag{3.24}$$

$$= \begin{pmatrix} 1 \\ 0 \\ 0 \\ 0 \end{pmatrix}. \tag{3.25}$$

Therefore, the application of the Grover iteration leads to an state in which the amplitude of the solution state has been amplified from $1/2$ to 1, which lead to a probability amplification from $1/4$ to 1. Measurement at this stage will result in the solution with probability 1. This step in Grover's algorithm is visualized in Figure 3.4.

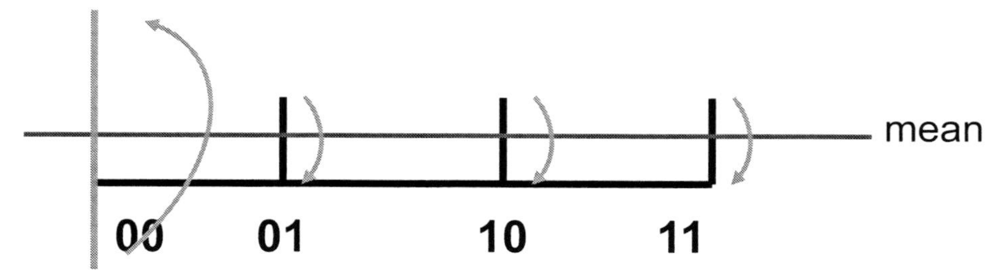

Figure 3.3: The inversion around the mean of the quantum amplitudes.

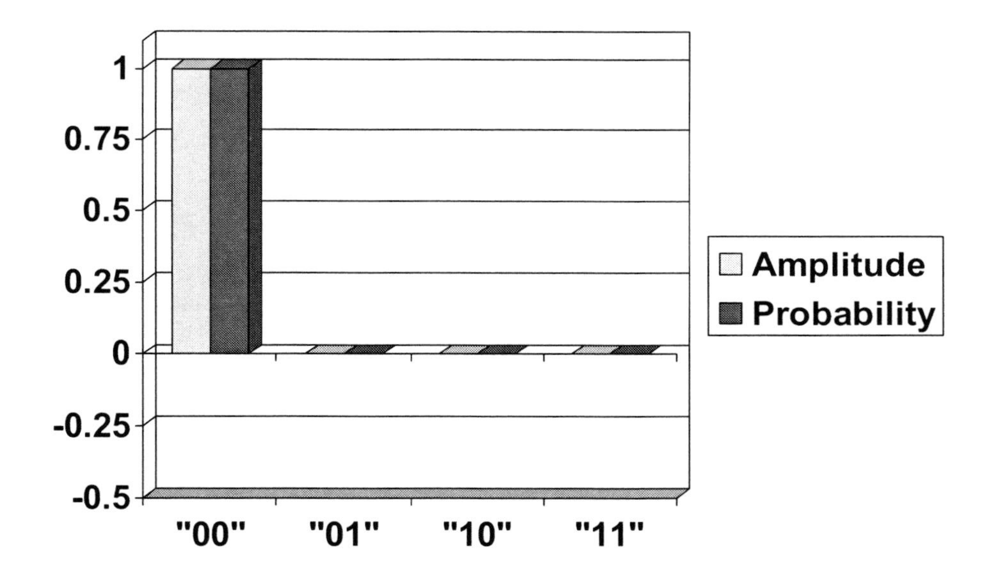

Figure 3.4: Visualization of the application of the inverse around the mean operator in Grover's Algorithm for 2 qubits, and "00" as the solution to the search problem. At this stage the solution has amplitude and probability equal to 1.

Therefore, in this simple example the correct solution is obtained with probability 1 after a single Grover iteration. What is important to note is that the solution was found using only one oracle computational step. In contrast, the classical solution would require up to four applications of the oracle to find the solution. For the case of $n > 2$ qubits, Grover's algorithm will require multiple iterations, but the number will be sublinear with N and is, thus, superior to the best possible classical alternative.

Let us estimate now how many Grover iterations need to be performed. To this end, we can rewrite the uniform superposition as the sum of the nonsolutions and the solution:

$$|\mathbf{\Psi(0)}\rangle = \alpha(0)\sum_{x\neq y}|\mathbf{x}\rangle + \beta(0)|\mathbf{y}\rangle \tag{3.26}$$

where y is the solution to the search problem.

Iteratively applying the Grover operator r times to the n-qubit uniform superposition state $|\mathbf{\Psi}\rangle$ gives the following. For the first iteration we have:

$$\begin{aligned} |\mathbf{\Psi(1)}\rangle &= G|\mathbf{\Psi(0)}\rangle &(3.27)\\ &= G\left(\frac{1}{\sqrt{N}}\sum_{i=0}^{N-1}|\mathbf{i}\rangle\right) &(3.28)\\ &= \alpha(1)\sum_{x\neq y}|\mathbf{x}\rangle + \beta(1)|\mathbf{y}\rangle\ . &(3.29)\end{aligned}$$

And for the rth iteration:

$$\begin{aligned} |\mathbf{\Psi(r)}\rangle &= G^r|\mathbf{\Psi(0)}\rangle &(3.30)\\ &= G^{r-1}|\mathbf{\Psi(1)}\rangle &(3.31)\\ &= \ldots &(3.32)\\ &= G|\mathbf{\Psi(r-1)}\rangle &(3.33)\\ &= \alpha(r)\sum_{x\neq y}|\mathbf{x}\rangle + \beta(r)|\mathbf{y}\rangle\ . &(3.34)\end{aligned}$$

Then, after r Grover iterations we obtain:

$$|\mathbf{\Psi(r)}\rangle = \alpha(r)\sum_{x\neq y}|\mathbf{x}\rangle + \beta(r)|\mathbf{y}\rangle \tag{3.35}$$

with the normalization condition:

$$|\alpha(r)|^2(N-1) + |\beta(r)|^2 = 1\ . \tag{3.36}$$

This result suggests the following convenient way to parametrize $\alpha(r)$ and $\beta(r)$:

$$\begin{aligned} \alpha(r) &= \frac{\cos((2r+1)\theta)}{\sqrt{N-1}} &(3.37)\\ \beta(r) &= \sin((2r+1)\theta) &(3.38)\end{aligned}$$

with

$$\sin\theta = \frac{1}{\sqrt{N}} \tag{3.39}$$

which clearly satisfies the initial condition:

$$\beta(0) = \sin(\theta) = \frac{1}{\sqrt{N}} . \tag{3.40}$$

Clearly, $|\beta(r)|^2$ is the probability of finding the solution to the search problem after r iterations. Therefore, to find the solution to the search problem with high probability we require:

$$\beta(r) \approx 1 \tag{3.41}$$
$$\alpha(r) \approx 0 . \tag{3.42}$$

Then:

$$\beta(r) = \sin((2r+1)\theta) \approx 1 \Rightarrow (2r+1)\theta \approx \frac{\pi}{2} . \tag{3.43}$$

And for large N:

$$\sin\theta = \frac{1}{\sqrt{N}} \ll 1 \Rightarrow \sin\theta \approx \theta \Rightarrow \theta \approx \frac{1}{\sqrt{N}} . \tag{3.44}$$

Therefore:

$$(2r+1)\theta \approx \frac{\pi}{2} \Rightarrow r \approx \frac{\pi}{4}\sqrt{N} - \frac{1}{2} . \tag{3.45}$$

As a consequence, for large N, the probability is maximized with $r \approx (\pi/4)\sqrt{N}$ iterations, which implies that *expected* $O\left(N^{1/2}\right)$ Grover iterations are required.[2] This sublinear $O\left(N^{1/2}\right)$ complexity provides a substantial improvement over the $O(N)$ complexity for a classical search.

Basically, Grover's algorithm performs inversions around the mean in such a way that the amplitudes of solution states increase while those of nonsolutions decrease. However, because all the operations involved in the algorithm are unitary, then a Grover iteration can be understood as a rotation in the space of quantum states.

The way in which the amplitudes change can be visualized geometrically is the following. Suppose we have the state:

$$|\boldsymbol{\Psi}(\boldsymbol{r})\rangle = \alpha(r) \sum_{x \neq y} |\boldsymbol{x}\rangle + \beta(r)|\boldsymbol{y}\rangle . \tag{3.46}$$

Then, after we apply a Grover iteration we get:

$$|\boldsymbol{\Psi}(\boldsymbol{r+1})\rangle = G|\boldsymbol{\Psi}(\boldsymbol{r})\rangle = \alpha(r+1) \sum_{x \neq y} |\boldsymbol{x}\rangle + \beta(r+1)|\boldsymbol{y}\rangle . \tag{3.47}$$

As shown in Figure 3.5 the net effect of a Grover iteration is the rotation of the original state in such a way that its projection to the axis of the solutions is increased.

[2] Big-Oh notation refers to worst-case complexity unless modified by an adjective such as "expected" (or followed by words such as "with high probability") when the bound is probabilistic in nature. In the quantum computing literature, however, it is common for the probabilistic qualifiers to be assumed implicitly. This is not necessarily a serious abuse of notation because the probability can be made arbitrarily high. For this reason, and for notational simplicity, we adopt the same convention.

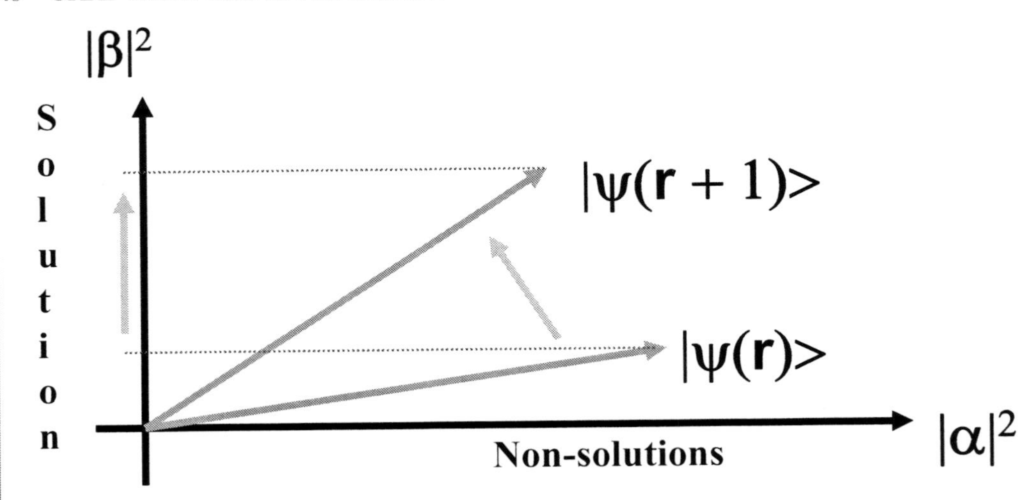

Figure 3.5: Visualization of the effect of a Grover iteration.

A common misunderstanding of Grover's algorithm is to expect that if we apply more iterations to the state, then we will obtain higher probability of success upon measurement. However, this is not the case: Grover's algorithm is not an iterative process that asymptotically approaches the solution.

This algorithmic property is due to the periodicity of the trigonometric functions in the parametrization of the state after a Grover iteration. Indeed, as the *sine* function increases its value, the probability of success increases. As we start with a small value for the argument, and as the argument increases, the value of the function increases. But at some point, an increase in the argument of the *sine* function will actually decrease its value. That is, more iterations than necessary will take the state farther away from the solution. This effect is illustrated Figure 3.6. Therefore, it is very important to apply a number of iterations that correspond to the ones specified by the algorithm.

As an example, let us consider the case of the search problem with a single solution using 3 qubits. Here, the search space has $2^3 = 8$ elements and we expect the highest probability after about 2 iterations. On an uniform superposition, before applying the algorithm, the probability of success is of about 0.125. Then, after a single Grover iteration the probability of success is augmented to 0.78. And after two iterations it becomes 0.95. However, after applying the Grover iteration a third time, the probability of success is reduced to 0.33.

In the above examples we have considered a search for a specified element, but Grover's algorithm can be applied to find the solution for any arbitrary oracle. This is important because many classes of queries can be satisfied in sublinear time using classical data structures. For general queries, however, a classical algorithm can do no better than an $O(N)$ sequential search of the dataset.

It can be shown that, under the conditions of an unstructured and unsorted database, Grover's algorithm is optimal [57]. That is, no other quantum or classical algorithm can do better than

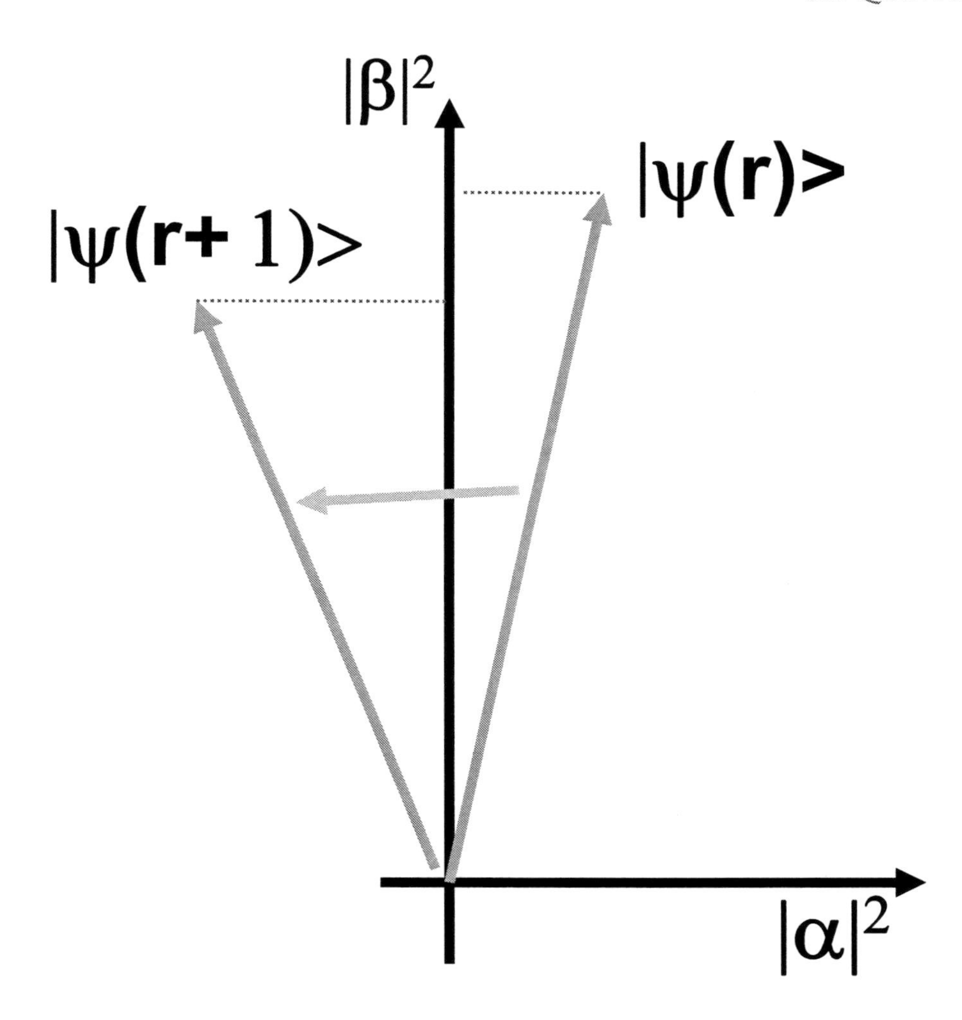

Figure 3.6: The effect of performing more iterations than those specified by Grover's algorithm.

$O(N^{1/2})$ under the assumptions of the problem (unstructured and unsorted dataset). However, although the $O\left(N^{1/2}\right)$ complexity of Grover's algorithm is quadratically more efficient than the best possible classical algorithm for general search problems, it is natural to wonder whether it can be made even more efficient. It turns out that it is provably optimal given the basic tenets of quantum physics.

It is important to remark again that the quantum complexity of $O\left(N^{1/2}\right)$ and the classical complexity of $O(N)$ for the problem of searching an unordered and unstructured database relate to the number of times we are required to use an oracle. That is, a classical algorithm requires $O(N)$

checks to the oracle, while Grover's algorithm requires $O\left(N^{1/2}\right)$ uses of the oracle. Clearly, Grover's algorithm is taking advantage of the intrinsic parallelism of the quantum computing model. The bottom line is that the optimality of Grover's algorithm refers to the number of times it requires to consult the oracle to produce a solution to the search problem, and nothing else.

As is the case with classical computing, however, the complexity of quantum search can be reduced substantially for *particular* classes of queries. This is particularly true for the case where the database is somewhat ordered or structured. This must be the case because classical computing is subsumed within the quantum model, and classical algorithm can satisfy a variety of queries in $O(\log(N))$ time using binary search, and it can satisfy exact-match queries in $O(1)$ expected time using hash tables.

3.1.4 GENERALIZED QUANTUM SEARCH

Amplitude amplification can be applied to more general classes of search problems in which a multiplicity of solutions may exist. In the case of k solutions, Grover's algorithm can find one solution in $O\left((N/k)^{1/2}\right)$ time – but only if the value of k is known. The need for the value of k can be seen from the formula for determining the number of iterations.

Indeed, after r iterations of the Grover operator the superposition with k solutions will look like:

$$|\boldsymbol{\Psi}(\boldsymbol{r})\rangle = \alpha(r) \sum_{non\text{-}solutions} |\boldsymbol{x}\rangle + \beta(r) \sum_{solutions} |\boldsymbol{y}\rangle \tag{3.48}$$

which normalization looks like:

$$|\alpha(r)|^2(N-k) + |\beta(r)|^2 k = 1 \tag{3.49}$$

This suggests the parametrization given by:

$$|\alpha(r)|^2(N-k) = \cos^2(2r+1)\theta \tag{3.50}$$

$$|\beta(r)|^2 k = \sin^2(2r+1)\theta \tag{3.51}$$

and therefore:

$$\beta(r) = \frac{1}{\sqrt{k}}\sin(2r+1)\theta \tag{3.52}$$

$$\alpha(r) = \frac{1}{\sqrt{N-k}}\cos(2r+1)\theta \tag{3.53}$$

with:

$$\sin\theta = \sqrt{\frac{k}{N}} \tag{3.54}$$

A before, high probability of measuring a solution implies:

$$\beta(r) \approx 1\sqrt{k} \tag{3.55}$$

$$\alpha(r) \approx 0 \tag{3.56}$$

And for large $N \gg k$:

$$\sin\theta = \sqrt{\frac{k}{N}} \ll 1 \Rightarrow \sin\theta \approx \theta \Rightarrow \theta \approx \sqrt{\frac{k}{N}} \tag{3.57}$$

Therefore:

$$(2r+1)\theta \approx \frac{\pi}{2} \Rightarrow r \approx \frac{\pi}{4}\sqrt{\frac{N}{k}} - \frac{1}{2} \tag{3.58}$$

As a consequence, for large N, the probability is maximized with $r \approx (\pi/4)\sqrt{N/k}$ iterations, which implies that $O\left((N/k)^{1/2}\right)$ Grover iterations are required.

In most practical applications k is unknown, which means that Grover's algorithm cannot be applied. It is possible to apply a *quantum counting* algorithm to determine the number of solutions. An optimal quantum counting algorithm has complexity $O\left((kN)^{1/2}\right)$, which would dominate the search complexity [10]. Fortunately, a different formulation of the amplitude amplification procedure can be applied to find a solution in $O\left((N/k)^{1/2}\right)$ time, even if k is unknown [9].

The fact that a solution can be found more efficiently, as k gets large, should be no surprise: as the number of solutions increases, the probability of sampling one of them increases. For example, if $k = O(N)$ we can find a solution in $O(1)$ time, just from random sampling. $O\left((N/k)^{1/2}\right)$ is provably optimal for this type of search problem; however, most practical applications require retrieval of all k solutions, not one of them chosen at random.

However, we cannot output the entire solution dataset using a single application of Grover's algorithm. Indeed, the superposition of states for the last iteration of Grover's algorithm, with known k, looks like:

$$|\boldsymbol{Q_A}\rangle = G^r|\boldsymbol{\Psi(0)}\rangle \approx \sin\left((2r+1)\theta\right)\frac{1}{\sqrt{k}}\sum_{solutions}|\boldsymbol{y}\rangle \tag{3.59}$$

where the probability of finding a nonsolution is presumed to be small and has been neglected in the equation. Thus, $|\boldsymbol{Q_A}\rangle$ is very close to be a superposition of solution states.

Unfortunately, a measurement to retrieve a solution causes the superposition to collapse, thus, losing all information relating to other possible solution states.

We could apply Grover's algorithm multiple times to retrieve the entire dataset of solutions. But to do so, we require about $O(k\log(k))$ applications of Grover's algorithm, and the overall complexity of the solution is $O(\sqrt{kN}\log(k))$. For a dense dataset, where k is comparable to N, the complexity is reduced to $O(N\log(N))$. This is a suboptimal solution, as the brute force classical search leads to an improved complexity of $O(N)$.

3.2 GROVER'S ALGORITHM WITH MULTIPLE SOLUTIONS

If it is known that there are exactly k solutions, the complexity of Grover's algorithm is $O\left((N/k)^{1/2}\right)$. In other words, if there are many solutions, Grover's algorithm can more efficiently find one of them. Grover's algorithm requires the value of k, and that value is rarely known a priori in practical situations, so it must be computed using a quantum counting algorithm. Unfortunately, the best possible (provably optimal) quantum counting algorithm has complexity $O\left([(k+1)(N-k+1)]^{1/2}\right) = O\left((kN)^{1/2}\right)$ [10].

As already mentioned, there is an alternative to Grover's algorithm that can search efficiently for a single solution without knowledge of k. Unfortunately, most practical applications require the retrieval of all the solutions, not just one of them. As will be discussed in this section, the lower bound for retrieving the k solutions is determined by the lower bound on counting them. Counting therefore can be done as a first step in the retrieval algorithm without undermining the overall complexity. Grover's algorithm can then be applied with known k to essentially sample one of the solutions at random. This can be repeated $O(k \log(k))$ times to ensure with high probability that all k solutions are sampled. This algorithm can be summarized as follows:

Suboptimal Quantum Multi-Object Search Algorithm

1. Perform a quantum counting operation using the oracle function f to determine the number k of solutions in dataset S of size N. Complexity: $O\left((kN)^{1/2}\right)$

2. While *counter* $< k$ do: (expected number of iterations: $O(k \log(k))$)

   (a) Initialize quantum register r to a uniform superposition of N indices corresponding to elements in S. Complexity: $O(1)$

   (b) Apply Grover's algorithm to sample one of the k solutions according to f in $O\left((N/k)^{1/2}\right)$ time.

   (c) Insert the sampled solution into a solution set (no duplicates) implemented with a hash table and increment *counter*. Complexity: $O(1)$

3. End Repeat. Complexity of loop: $O\left((kN)^{1/2} \log(k)\right)$

The overall complexity is dominated by the $O(k \log(k))$ applications of Grover's algorithm, each of which has $O\left((N/k)^{1/2}\right)$ complexity.

The complexity to retrieve all the solutions is clearly $\Omega(N)$ when $k = N$. Classical exhaustive search therefore has optimal worst-case complexity. However, the above quantum algorithm has complexity $O(N \log(N))$ in the worst case and is therefore suboptimal. The problematic logarithmic factor results from the repeated sampling of already-known solutions. This can be avoided if a different oracle is applied for each iteration so that previously-identified solutions are not resampled during subsequent iterations [35]. This can be accomplished by augmenting the elements of the indexed set S with a bit vector which is used to mark solutions as they are sampled. Thus, the augmented oracle only regards an index as a solution if its corresponding element in S is unmarked and is a solution to the original oracle. The new algorithm can be summarized as follows:

Optimal Quantum Multi-Object Search (QMOS) Algorithm

1. Perform a quantum counting operation using the oracle function f to determine the number k of solutions in dataset S of size N. Complexity: $O\left((kN)^{1/2}\right)$
2. While $k > 0$:
   (a) Initialize quantum register r to a uniform superposition of N indices corresponding to elements in S. Complexity: $O(1)$
   (b) Apply Grover's algorithm to sample one of the k solutions according to f in $O\left((N/k)^{1/2}\right)$ time and add it to the output dataset in $O(1)$ time.
   (c) Augment the oracle function f so that the sampled solution is marked as a nonsolution. Complexity: $O(1)$
   (d) $k \longleftarrow k-1$
3. End While. Complexity of loop: $O\left((kN)^{1/2}\right)$
4. Use the solution set to unmark all the elements of S in the oracle so that the complexity of future queries is not compromised by an $O(N)$ initialization step. Complexity: $O(k)$

The overall complexity is determined by the claimed complexity of $O\left((kN)^{1/2}\right)$ for the loop in step $2b$. This iteration can be expressed as the sum

$$\sum_{i=1}^{k}(N/i)^{1/2} = N^{1/2}\sum_{i=1}^{k} i^{-1/2} . \tag{3.60}$$

Using the well-known result

$$\sum_{i=1}^{n} i^{c} \in O\left(n^{c+1}\right) \text{ for real } c \text{ greater than } -1 , \tag{3.61}$$

we obtain the complexity $N^{1/2}O(k^{1/2}) = O\left((kN)^{1/2}\right)$ as claimed. Furthermore, this overall complexity is optimal by virtue of the optimality of the same complexity for quantum counting. In other words, retrieving the k solutions yields the value of k and therefore cannot be accomplished with complexity better than that of the optimal counting algorithm.

The general quantum search algorithm, QMOS, is directly applicable to the general-purpose database problem in which it is desirable to support arbitrary queries with sublinear complexity. Traditional databases can only provide sublinear complexity for predefined classes of queries and so cannot efficiently support many types of data mining and analysis applications. In this case, quantum searching provides a complexity improvement over classical search only by virtue of its generality. There are also important special classes of queries, such as, multi-dimensional range searching, for which quantum search is more efficient than the best possible classical search algorithm. These and other applications will be discussed in the following chapter.

3.3 FURTHER APPLICATIONS OF AMPLITUDE AMPLIFICATION

Quantum amplitude amplification algorithms are not restricted to the problem of searching solutions from an unsorted and unstructured database. To name just a couple, these methods can also be used to perform a count on the number of solutions, and to find the minimum, maximum and mean of a dataset [15].

From its original inception, Grover's algorithm was advertised as a tool to perform efficient searches of items inside a large database. As we discussed, Grover's algorithm can be proved to be optimal if the database is unsorted and unordered. That is, if no structure is imposed or created on its elements, then brute force is the only viable alternative in the classical domain. In this case, the quantum solution performs quadratically better than the classical solution. However, if you think about it, most databases of scientific, industrial, or financial interest are made of alphanumerical strings which can be sorted and structured to conduct fast exact match searches. Even abstract concepts, such as, colors or geometric figures can be structured one way or another (using RGB values to represent color and a mesh of polygons to represent figures).

Thus, for most practical applications a clever software developer will organize the database before conducting a series of searches. By using data structures, such as, binary trees or Hash tables, the developer may write a program that retrieves arbitrary elements from the database in constant or logarithmic time. Under this circumstances, classical methods will outperform Grover's algorithm. However, as we will see in the next chapter, there is a wide class of computational problems that can benefit from Amplitude Amplification algorithms.

3.4 SUMMARY

Amplitude Amplification is a powerful technique for the development of quantum algorithms. For the specific case of Grover's algorithm, searching for an item in an unordered and unsorted database can be performed quadratically faster than in classical computing. Similarly, the QMOS algorithm uses amplitude amplification to extract the entire solution set from an unordered and unsorted database. At this point it is important to remark that Grover's algorithm has been proved to be optimal within the assumptions of an unsorted and unstructured database. Indeed, if the elements of the dataset can be ordered using a data structure, then a classical computer could perform the search exponentially faster than Grover's algorithm.

CHAPTER 4

Case Study: Computational Geometry

Computational geometry [38, 41] is concerned with the computational complexity of geometric problems that arise in a variety of disciplines such as:

- Computer Graphics
- Computer Vision
- Virtual Reality
- Augmented Reality
- Multi-Object Simulation and Visualization
- Multiple-Target Tracking

Many of the most fundamental problems in computational geometry involve multi-dimensional searching, i.e., searching for objects in space that satisfy certain query criteria [38, 41]. Efficiently supporting such queries typically involves the use of data structures for accessing specified objects and, also, for representing those objects. For example, the spatial extent of an object may be represented in the form of a convex hull[1].

Virtually, all of the computational bottlenecks in computer graphics for video games or visual effects involve some variant of spatial search [18, 19]:

1. Collision detection involves the identification of intersections among spatial objects.
2. Ray tracing involves the identification of the spatial object that is intersected first along a given line-of-sight.

In the most general case, the spatial objects in computational geometry problems have an arbitrary shape and size. As such, these problems may be very difficult to solve with an algorithm that performs substantially better than exhaustive search. For instance, in Figure 4.1 we can appreciate how difficult it is to establish which objects intersect other objects if all of them have arbitrary shape and size. In this case, we practically have to determine if any point in object i intersects any point in object j.

[1] Let us recall that the convex hull of a set of points X is the smallest convex envelope containing all elements of X.

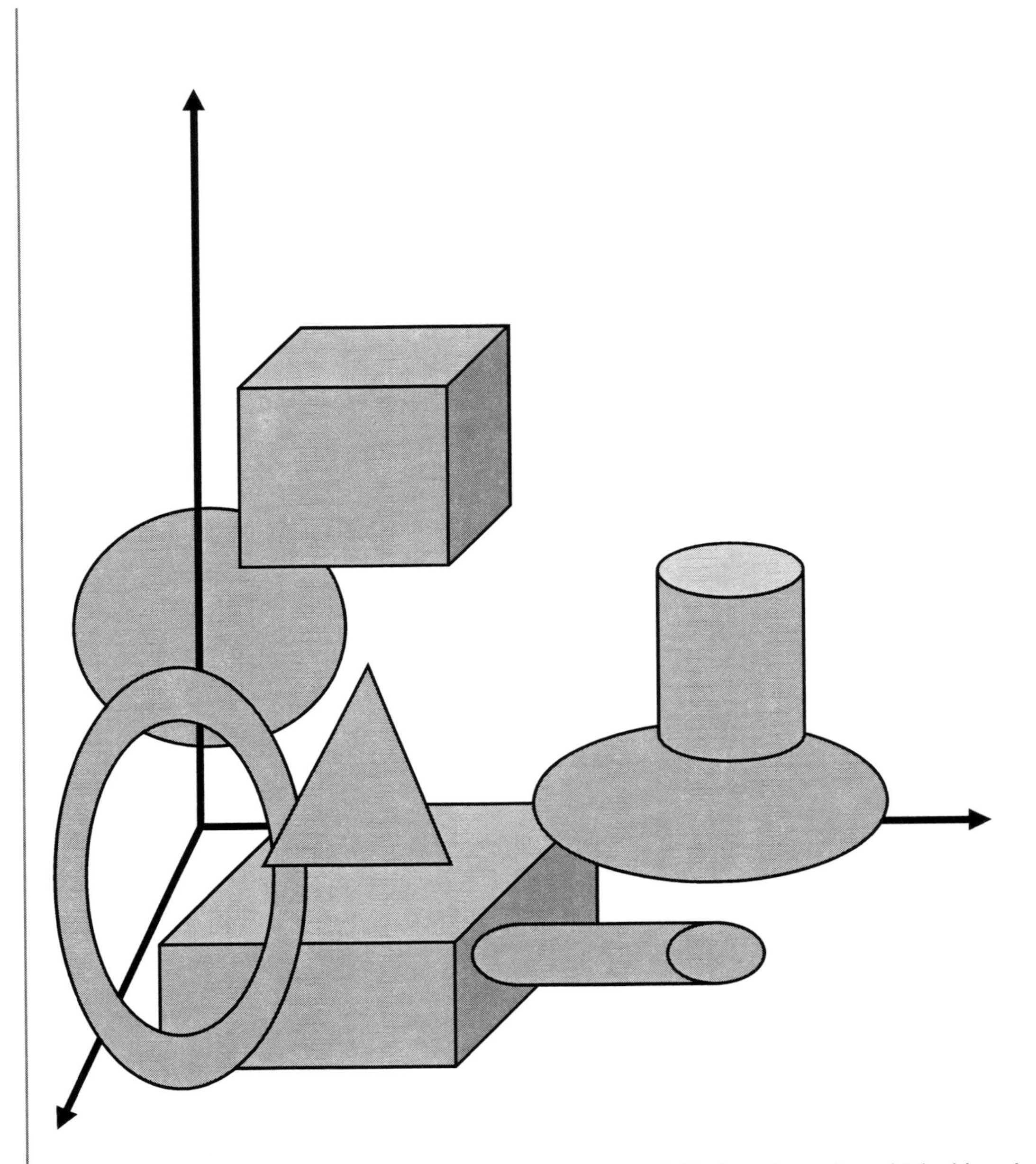

Figure 4.1: For objects of arbitrary shape and size, it is very difficult to determine which objects intersect other objects. Most algorithms will not perform much better than exhaustive search.

On the other hand, the problem is greatly simplified if the objects are simple coordinate-aligned boxes. (See Figure 4.2.) Indeed, in this case it is enough to check if the projection of any of the edges of box 1 does not overlap the respective edge projection of box 2, in which case the objects do not intersect. Therefore, a common technique to find intersections between spatial objects usually involves as a first step to wrap each object in a tight, coordinate-aligned bounding box. And then, the full intersection computation is performed only if the bounding boxes intersect each other.

Therefore, the most efficient known algorithms for computational geometry problems are applicable to only very special cases, e.g., involving coordinate-aligned boxes. Theoretical bounds for any classical algorithm to address more general cases are very pessimistic with regard to prospects for large-scale, real-time applications. Of course, there is no classical alternative to performing an exhaustive search when the types of queries are not known a priori.

The advantage of quantum computing, as discussed in the previous chapter, is that it offers rootic complexity for all types of queries. In this chapter we will discuss more specific classes of search problems relevant to computer graphics, simulation, and visualization [33, 36, 31, 32]. In particular, we will describe how to implement quantum algorithms to efficiently solve a variety of computational geometry problems that appear in:

- General Spatial Search Problems
  - Multi-dimensional Search and Intersection Queries
  - Convex Hull Determination
  - Object-Object Intersection Identification
  - All-Pairs Intersection Identification
- Computer Graphics
  - Z-Buffering
  - Ray Tracing
  - Radiosity
  - Management of Level of Detail

4.1 GENERAL SPATIAL SEARCH PROBLEMS

CC algorithms for satisfying range queries, interval intersection queries, and many other types of one-dimensional retrieval operations achieve $O(\log(N))$ complexity by using some variant of binary search. Unfortunately, multi-dimensional generalizations that are constrained to using minimal $O(N)$ space, such as, the multi-dimensional binary search tree (also known as the k-d tree), provide only rootic query time that depends on the number of dimensions [41]. Specifically, orthogonal range queries in d dimensions require $O(N^{1-1/d})$ time ($O(N^{2/3})$ in three dimensions) in the worst

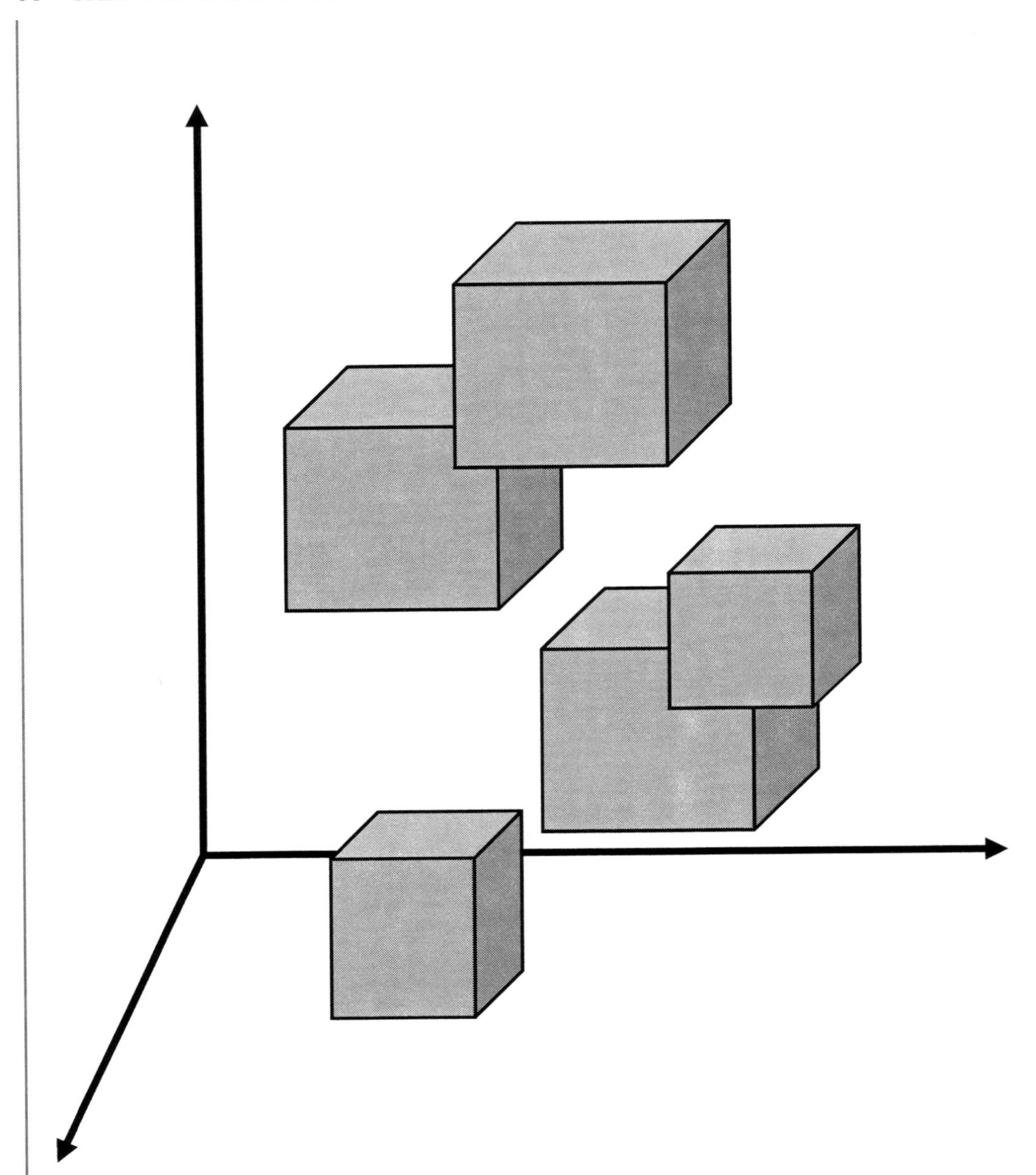

Figure 4.2: Finding which objects intersect each other is greatly simplified if the objects in the database are coordinate-aligned boxes.

case using a multi-dimensional Binary Search Tree (BST). Obviously, this represents a lower bound for more general types of queries involving non-orthogonal objects.

In most practical applications involving the retrieval of geometrically proximate objects (sometimes referred to as "near neighbors"), the value of k is $O(1)$ (where k is the number of solutions to the search problem). In such cases, the complexity of QMOS is superior to that of a multi-dimensional BST for orthogonal range queries in all dimensions greater than two. The advantage of the QMOS algorithm in 4D simulations, in which time represents an additional dimension, is even more substantial as the CC complexity for general point location has a lower bound of $\Omega(N^{3/4})$, and no CC algorithm has been found that achieves this lower bound. In fact, it may be the case that $O(N^{1-1/d})$ cannot be achieved in minimal $O(N)$ space except in the special case of orthogonal query objects.

Recent analysis [6] suggests that the best CC spatial query algorithms in 2D can only achieve polylogarithmic query time, i.e., $O(\log^c(N))$ for some constant $c > 1$, at the expense of $\Omega(N \log N)$ space. It is likely that CC data structures for most types of spatial queries will have much higher space complexities (with logarithmic factors that are exponential in d), but even this lower bound complexity suggests that space will become an obstacle for large values of N. Indeed, for a typical scenario in computer graphics, if the space complexity is $O(N \log^d N)$ and $N = 10^7$, then the space overhead in three spatial dimensions ($d = 3$) is a multiplicative factor close to 13000.

Because the CC algorithms that achieve query-time complexity superior to that of QMOS do so at the expense of nonlinear space, the $O\left((kN)^{1/2}\right)$ query complexity and $O(N)$ space complexity of QMOS are very attractive. In addition, the QMOS solution is valid for *general* objects, and it is not restricted to coordinate-aligned boxes.

Table 4.1 summarizes these results and offers a comparison of the resulting complexities when using different multi-dimensional spatial search methods. In this table, the pre-processing time refers to the number of operations necessary to move the dataset from disk to memory and to build the data structure.

It is also important to note that the QMOS has space complexity of $O(N)$, even though we argued that the quantum computing model has an exponentially large computational space. The reason is that the dataset has to reside on a classical memory, so it can be removed and queried multiple times (recall that a query to a quantum memory would collapse the quantum state of the device).

This complexity analysis for QMOS holds both for point retrieval or for finding all objects which intersect a given query object as long as identifying whether two objects intersect can be done in $O(1)$ time. If not, the complexity to find the k intersections must be multiplied by the complexity to determine whether a given pair of objects intersect. Both aspects of the problem are considered in the following two subsections.

These results can be better appreciated by plotting the behavior of the query time and space resources for large values of N. For the sake of illustration, let us consider the simple case where we have a single solution, $k = 1$, in four dimensions, $d = 4$. Furthermore, let us suppose that the proportionality constant of the complexities is just a factor of 1. With these assumptions, Figure 4.3

Table 4.1: Complexity comparison of classical [41] and quantum algorithms for satisfying multi-dimensional spatial search and intersection queries with solution size k in d dimensions. (Box queries refer to the identification of points or orthogonal boxes that are in or intersect a given orthogonal box. General queries refer to objects other than boxes for which the intersection between a pair of objects can be identified in $O(1)$ time.)

Search Type	Preprocessing Time	Query Time	Space Resources
Classical for general queries	$O(N)$	$O(N)$	$O(N)$
Classical for box queries	$O(N \log N)$	$O(N^{1-1/d} + k)$	$O(N)$
Classical for box queries	$O(N \log^{d-1} N)$	$O(\log^d(N) + k)$	$O(N \log^{d-1} N)$
QMOS for general queries	$O(N)$	$O\left((kN)^{1/2}\right)$	$O(N)$

shows the behavior of the query time. As can be observed, for $N = 10^{20}$, the quantum solution ($QC = 10^{10}$) is significantly faster than the classical for general objects ($CC = 10^{20}$) and the classical for box queries using linear space ($LS = 10^{15}$).

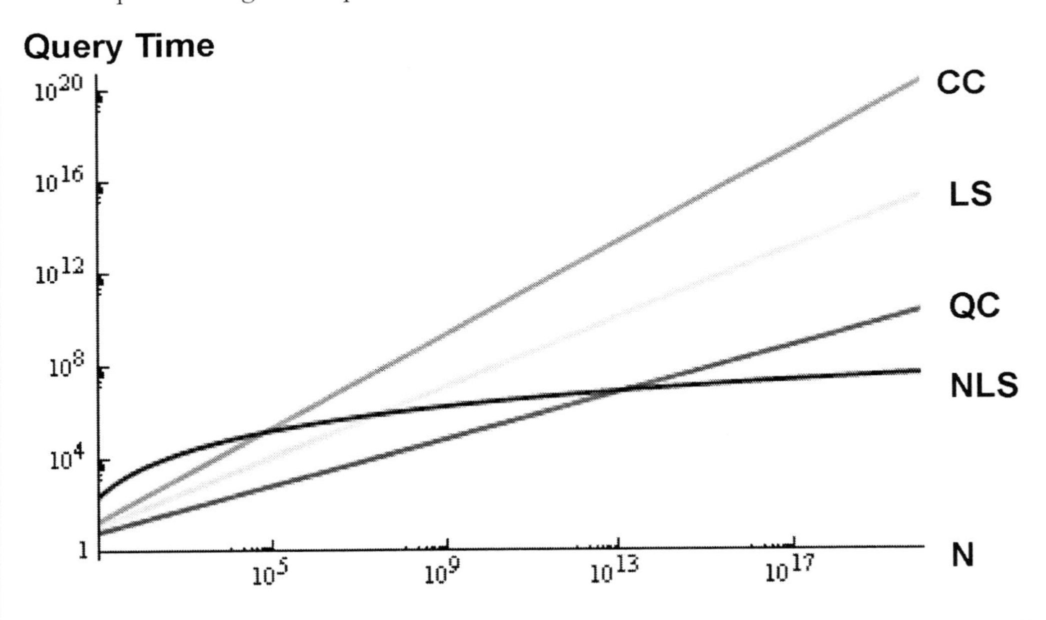

Figure 4.3: Query time for several multi-dimensional search algorithms: linear space classical for general queries (CC), linear space classical for box queries (LS), non-linear space classical for box queries (NLS), and quantum for general queries (QC). In this example we have assumed a single solution, $k = 1$, four dimensions, $d = 4$, and a proportionality factor for the complexities equal to one.

On the other hand, for $N = 10^{20}$ the quantum solution is slower than the classical solution for box queries using non-linear space ($NLS = 10^7$). However, if we look at the amount of space required, shown in Figure 4.4, we observe that for $N = 10^{20}$ the classical solution for box queries using non-linear space requires about 100, 000 times more memory space than the other three cases that use linear space. Clearly, the large memory requirement for non-linear space classical solutions is a serious limitation to the practical implementation of these methods.

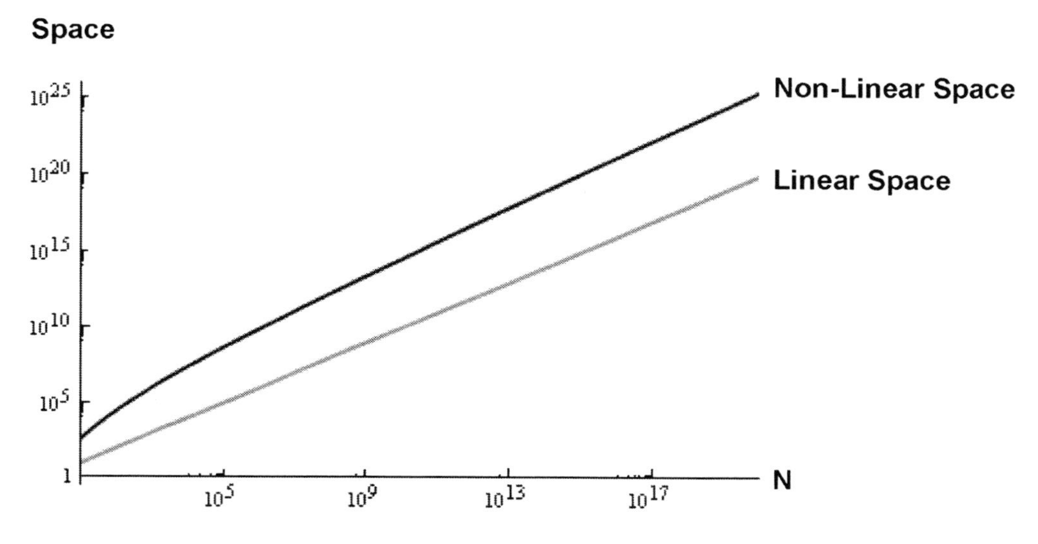

Figure 4.4: Space requirements for linear and non-linear space multi-dimensional search algorithms. In this example we have assumed a single solution, $k = 1$, four dimensions, $d = 4$, and a proportionality factor for the complexities equal to one.

Furthermore, the advantage in query time provided by the non-linear solution happens for a relatively large value of N. As seen in Figure 4.3, the transition occurs at around $N = 10^{13}$. That is, non-linear space algorithms for a particular application require N to be big enough for the better query time complexity to become evident. But in this case, the algorithm also requires a very large amount of storage, a situation that may render it unfeasible.

As a consequence, most practical applications of multi-dimensional searches involve implementations that require linear space storage. In any event, the advantages of a possible tradeoff between query time and space will depend on the particulars of the problem at hand and the amount of computational resources available to the user. Therefore, the quantum solution becomes an attractive alternative because of its efficient query time combined with simple linear space requirements.

4.1.1 QMOS FOR OBJECT-OBJECT INTERSECTION IDENTIFICATION

When simulating the motion of objects in a multi-object simulation or Virtual Reality (VR) system, it is necessary to identify the interaction of any pair of objects so that their respective states can be updated, e.g., to reflect the kinematic changes resulting from a collision. Collision detection introduces two computational challenges. The first involves the determination of whether two given objects are interacting as indicated by an intersection of their surfaces. This can be computationally expensive for objects whose surfaces are non-convex. The second challenge is to identify which pairs have interacted out of a set of N moving objects. This is often the most computationally expensive part of a simulation or VR system because $O(N^2)$ time is required to check every pair, and the most commonly used algorithms for reducing this complexity tend to be data-dependent with scaling between $O(N \log(N))$ and $O(N^2)$. Grid-based algorithms (possibly using hash indexing) can make time/space tradeoffs in which storage is proportional the volume of the search space or the query time is proportional to the volume of the objects. However, none of these approaches appear to be practical for very large problems.

The most basic capability necessary for collision detection is to determine whether two given objects intersect, e.g., after their states have been motion-updated during a timestep. If the objects are convex, then intersections can be performed very efficiently. However, if they are very complex dynamic objects, e.g., streamers blowing in the wind, then determining intersection can be computationally expensive.

A standard mechanism for reducing the computational overhead of intersection detection is to apply coarse gating criteria that can efficiently identify pairs that do *not* intersect. One such mechanism is the use of bounding volumes, such as, spheres, coordinate-aligned boxes, and convex hulls. If the bounding volumes for two given objects do not intersect, then the objects cannot possibly intersect, so there is no need to perform more complex calculations involving the actual objects themselves.

One approach for gating is to compute the 2D convex hulls of each 3D object obtained by projecting (implicitly) its surface points onto each coordinate plane. With these hulls it is possible to determine that two given objects do not intersect simply by showing that their 2D convex hulls do not intersect in one of the coordinate planes. No projection operation needs to be explicitly performed because the projection of a point onto one of a coordinate plane is found directly from two of the point's three coordinates, so the computation of the convex hulls represent the main computational expense.

There are many efficient classical algorithms for computing 2D convex hulls, but the most efficient of these algorithms requires $\Omega(N \log(h))$ time for N objects with h points forming the convex hull [41]. The relevant question is whether a better QC algorithm exists. One approach for finding such an algorithm is to examine each of the known CC algorithms to determine whether Amplitude Amplification Algorithms can be productively applied. It turns out that one classical algorithm, the Jarvis March [41], can exploit a quantum solution.

The Jarvis March begins by identifying one point on the convex hull. This can be achieved by finding the point with the minimum x-coordinate value. The next point, in clockwise order, on the convex hull can be found by computing the angles between the line $y = 0$ through the first point and the lines determined by the first point and every other point in the dataset. The line having the smallest angle (measured clockwise) goes through the next point on the convex hull. The same procedure is repeated using the line through the second point. Thus, each point on the hull is found successively in $O(N)$ time for an overall complexity of $O(Nh)$.

The attractive feature of the Jarvis March algorithm, from a QC perspective, is that each successive point can be determined after the application of a simple calculation for each of the points in the dataset. Specifically, the angles for the points in each step can be computed, and the minimum point retrieved, in $O\left(N^{1/2}\right)$ time using Grover's Algorithm. This reduces the complexity of the overall convex hull determination to $O(N^{1/2}h)$, i.e., sublinear time, if h is a constant, which is often the case in practice.

The above complexity can be further improved through the application of Akl-Toussaint heuristics with QMOS[2]. Specifically, a set of $O(1)$ extreme points (e.g., ones with minimum or maximum x and/or y coordinates) can be found using Grover's algorithm in $O\left(N^{1/2}\right)$ time. Under assumptions which hold in a variety of practical contexts, the number of points from the dataset, which do not fall within the convex hull of the extreme points, is expected to be $O(h)$. QMOS can be used to find this $O(h)$ superset of points (followed by an $O(h\log(h))$ step to extract the actual hull) in $O\left((hN)^{1/2}\right)$ time, which is better than the above Jarvis-based algorithm by a factor of $h^{1/2}$.

Table 4.2: Complexity comparison of the best classical 2D convex hull algorithm with that of two quantum alternatives. (N is the total number of points, and h is the number of points comprising the hull.)

Method	Computation Time	Space Resources
Classical (optimal)	$O(N\log(h))$	$O(N)$
Quantum (Jarvis w/ Grover)	$O(N^{1/2}h)$	$O(N)$
Quantum (Akl-Toussaint w/ QMOS)	$O((Nh)^{1/2})$	$O(N)$

Table 4.2 shows that the quantum 2D convex hull algorithm is considerably more efficient than the best classical algorithm as long as h is not large compared to N.

4.1.2 QMOS FOR BATCH INTERSECTION IDENTIFICATION

In the previous section we were concerned with the efficient determination of whether two objects, comprised of $O(N)$ surface points, intersect. In this section we are concerned with the identification of all pairs of intersecting objects in a dataset of N objects. This type of operation is necessary to identify all interactions or collisions in multi-object simulations.

The identification of all intersections can be accomplished in an online setting by applying the QMOS algorithm sequentially for each of the N objects with an oracle that identifies all objects which intersect the object. The resulting complexity is:

$$N * O\left((k_{\text{avg}}N)^{1/2}\right) = O\left(N^{1.5}k_{\text{avg}}^{1/2}\right) \tag{4.1}$$

[2]The Akl-Toussaint heuristics are methods used to accelerate the performance of convex hull algorithms. The basic idea is to try to get rid of those points that will not contribute to the convex hull as early as possible in the computation process.

where k_{avg} is the average number of intersections per object.

A more direct batch algorithm for finding the m total intersections from a set of N objects is possible by further exploiting the power of QC by doubling the size of the quantum register and constructing the Cartesian product of the N objects. In QC, this can be done in $O(1)$ time. QMOS can then be applied to the N^2 pairs to extract the m intersecting pairs in $O\left((N^2 m)^{1/2}\right) = O(N m^{1/2})$ time. For $m = N * k_{avg}$, this complexity is the same as the earlier result obtained by performing N sequential intersection queries; however, the overhead of iteration in the latter approach is completely avoided, i.e., it may be speculated that the batch identification of m intersections should prove to be significantly faster in practice.

To conclude, we have derived a general quantum algorithm for identifying all of the m intersecting pairs of objects in a dataset of size N. This algorithm has complexity $O(N * \sqrt{m})$, which matches the classical worst-case optimal algorithm when m is $O(N^2)$ and improves over the best possible classical algorithm in all other cases. In particular, when m is $O(N)$, which is the case in many practical applications in which each object intersects only a constant number of other objects, the complexity of the new quantum search algorithm is $O(N^{1.5})$. Although classical algorithms can achieve subquadratic complexity for special classes of objects, e.g., orthogonal boxes or convex polyhedra, our quantum algorithm achieves its complexity for any class of objects.

4.2 QUANTUM RENDERING

In computer graphics, each object that needs to be rendered is typically decomposed into several hundreds or thousands of smaller polygons or other surfaces [18]. These polygons (or surfaces) are stored in a database that is often very large in practical applications, e.g., at least a few million elements depending on the complexity of the scene. In this context, a scene is simply a collection of many objects. To render a scene, the visualization system applies a variety of rendering algorithms on each individual element of the database. These rendering algorithms have the common feature of performing several searches over the entire database of objects in a scene. Considering the large number of elements in the database, the search process tends to be the most significant bottleneck in the rendering pipeline. Therefore, these search operations are prime candidates to be modified and optimized using quantum amplification algorithms.

The question of whether quantum rendering algorithms can be defined at all was first considered by Andrew Glassner [20, 21, 22]. In the following subsections we provide QC algorithms for several important rendering and scene management applications [33, 36]. The results of our analysis include QC algorithms with superior worst-case complexity compared to their CC counterparts[3].

4.2.1 Z-BUFFERING

One of the simplest methods used to determine what polygons are visible in a scene that needs to be rendered is the Z-Buffering algorithm [18]. This is a very simple, but effective, algorithm. The first

[3]These quantum algorithms are modified, corrected, and improved versions of those presented by the authors at the *Siggraph* 2005 conference [33].

step is to scan-convert each polygon in the scene. This means that for each polygon, we calculate its projection onto the screen and determine what pixels in the screen need to be shaded according to the color of the polygon. Then, for each shaded pixel, we record the distance to the polygon. The algorithm iterates over each polygon in the database, and if for a given pixel there is a new polygon that is closer, we update the color of the pixel to represent the new polygon. In other words, for each polygon we draw its image on the screen and keep a buffer of values that represent its distance to the screen. If the polygon is the closest to the screen, then the respective pixels are shaded with the color of such a polygon (Figure 4.5). Therefore, the entire operation requires $O(pN)$ time, where p is average number of pixels in the projection area per scan-converted polygon. In other words, the problem is solved in $O(p)$ time per polygon.

This is not a search problem per se, but it can be modified in such a way that we can solve it using a variant of Grover's algorithm. This variant is a Grover's based algorithm used to find the element with the minimum value in a dataset [15]. In a sense, this problem is rather similar to some graph problems that have been studied within the context of quantum computing [40].

Let us suppose that for each pixel we create a quantum superposition of states, where each element of the superposition points at each of the N polygons in the database.

$$|\psi\rangle = \frac{1}{\sqrt{N}} \left(|P_1\rangle + |P_2\rangle + \ldots + |P_N\rangle \right) . \tag{4.2}$$

Now, for all the polygons P_i in the scene, we can calculate the values Z_i, the scan converted distance between the pixel ψ and polygon i. This step is performed in $O(1)$ steps, exploiting the natural parallelism of quantum computing. Then, for this pixel, we just need to determine the polygon with the smallest distance.

From here it is straightforward to apply the variant of Grover's algorithm to find the polygon with the minimum distance to the screen, which gives the polygon that needs to be used to shade the pixel. This process has to be repeated once for each pixel. Therefore, the complexity of the quantum Z-Buffering algorithm is $O\left(N^{1/2}\right)$ per pixel for an overall complexity of $O(PN^{1/2})$, where P is the total number of pixels. Although the factor of P is very large in the QC complexity, the QC algorithm scales much better with the number of polygons N than does the corresponding CC algorithm. Thus, as N grows into the billions, the QC algorithm achieves better query time complexity than the brute force CC algorithm. This is the scale of problem that is relevant to many virtual reality applications of interest, e.g., involving a complete virtual representation of an urban environment. Finally, one could argue that the use of a classical data structure could be used to speed up the classical algorithm. However, as we saw before, the best CC solutions are possible only at the expense of larger space complexity. Equally important, the quantum solution is for objects of arbitrary shape, and it is not restricted to simple polygons.

4.2.2 RAY TRACING

Ray tracing is one of the most commonly used rendering techniques [18, 19]. A ray tracer determines the visibility of surfaces by tracing imaginary rays from the viewer's eye to the objects in the scene.

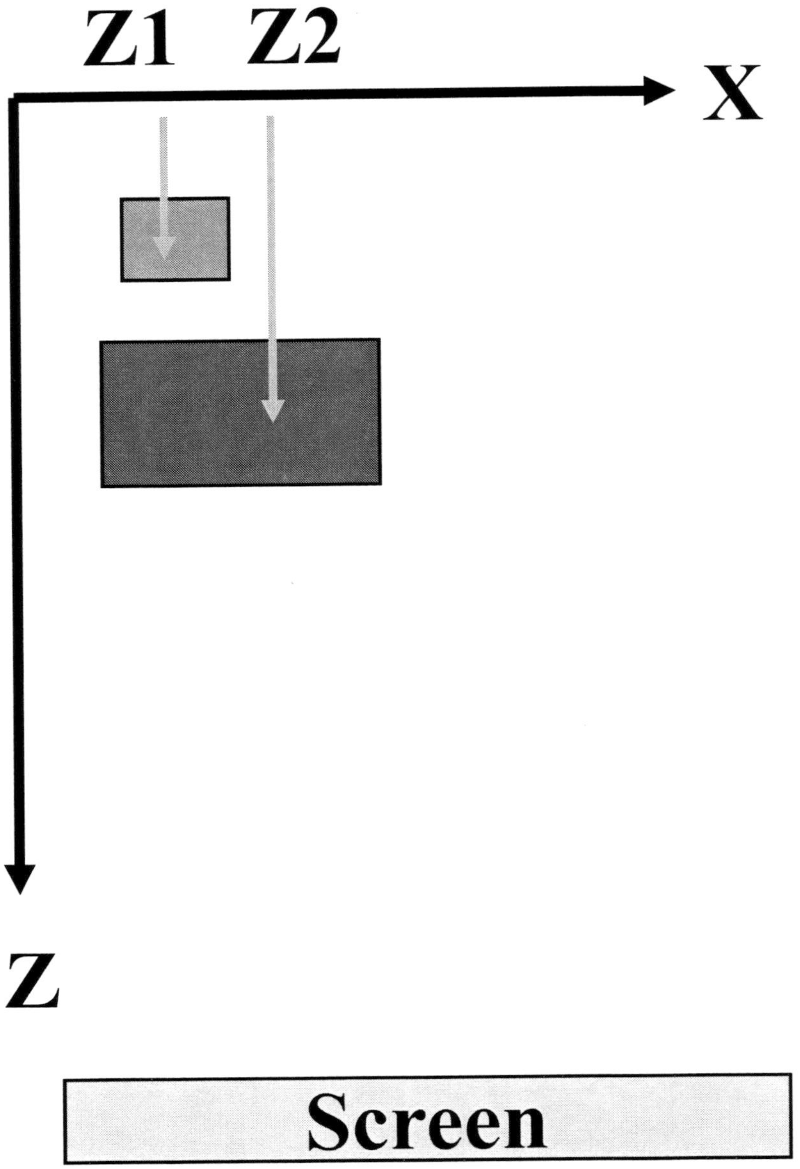

Figure 4.5: The Z-Buffering algorithm renders the object closer to the screen.

The intersection between a ray and an object determines the shade of the pixel, as seen in Figure 4.6. This process has to be performed for each pixel on the screen. Therefore, for each pixel on the screen, the ray tracer determines if there is an intersection between the ray and any object in the scene. If the ray intersects many objects, only the one closest to the screen is rendered. Thus, the ray tracer requires $O(N)$ operations per ray, as the intersections need to be determined for each of the N objects in the scene.

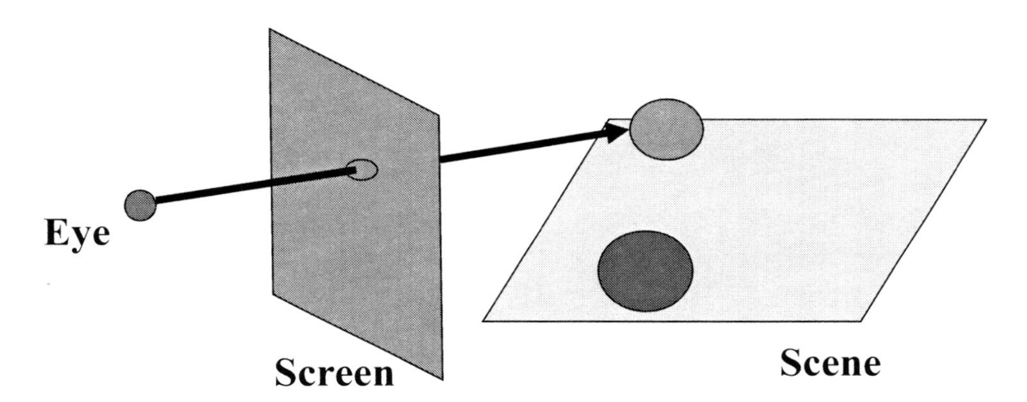

Figure 4.6: The ray-tracing algorithm calculates the intersection between imaginary rays emitted by the eye and the objects in the scene.

Although a variety of CC data structures exhibit much better performance than what the worst-case complexity might indicate, most of the data structures have been primarily analyzed empirically on special types of datasets of spatially proximate objects with small level-of-detail variance, e.g., objects in a single room. There is reason to believe that their performance for scenes involving views of large areas, e.g., through the trees and foliage of a virtual forest, will be substantially worse.

A ray tracer can be implemented in such a way that the objects in the scene reflect the original ray several times (Figure 4.7). This method allows the ray tracer to realistically model reflected and refracted light. However, each new ray in the iteration requires $O(N)$ further steps. So, if we consider a ray tracer with two secondary rays, as the one shown in Figure 4.7, the rendering time triples.

Because of the ray-tracing algorithm's intrinsic searching nature, which requires a search for intersections between a ray and a collection of N objects, ray tracing is a prime candidate to be optimized by means of quantum amplification algorithms. Therefore, the quantum ray tracer can be simply implemented to calculate the intersection between rays and polygons. And once we determine the polygon being intersected, we can perform further refinements to properly shade the corresponding pixel. The intersection between a ray and a polygon is a very simple procedure, and it involves basic analytical geometry to calculate the intersection.

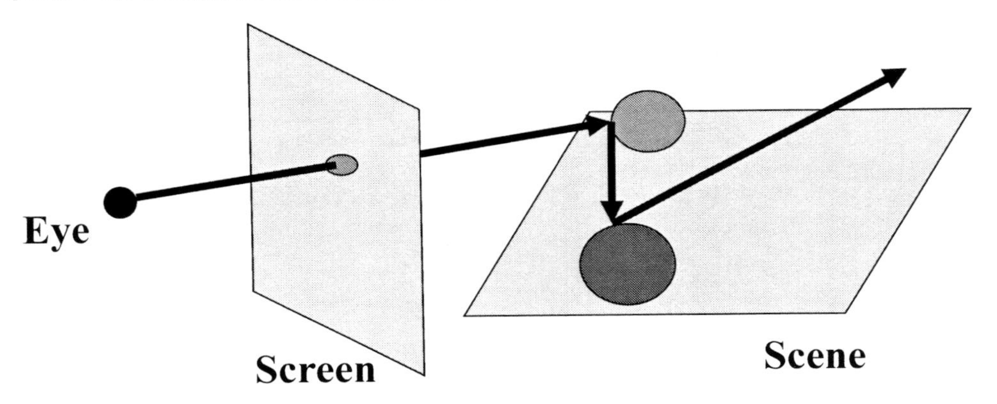

Figure 4.7: Ray tracing with multiple reflections.

To implement a quantum ray tracer, we create a state $|\Psi>$ that encodes all the polygons in the scene:

$$|\Psi> \quad = \quad \frac{1}{\sqrt{N}}(|\mathbf{00...00}\rangle + |\mathbf{00...01}\rangle + |\mathbf{00...10}\rangle \; + \; ...) \tag{4.3}$$

where each element in the superposition "marks" or "points at" one of the different polygons in the scene. This superposition of states is used for each ray in the ray tracing process. We also need to define an oracle function f that will act on each state of the superposition and it is such that:

$$f(x) \quad = \quad 1 \text{ if x intersects the ray} \tag{4.4}$$
$$f(x) \quad = \quad 0 \text{ any other case .} \tag{4.5}$$

After adding some ancillary qubits, we can use quantum parallelism to evaluate $f(x)$ for each element of the superposition (each object in the scene) in a single step. From here it is straightforward to use QMOS to determine all the k polygons that intersect a ray in $O((Nk)^{1/2})$. Then, we can use the variant of Grover's algorithm to find the minimum in $O(k^{1/2})$ time. The overall complexity for the quantum ray tracing algorithm becomes $O((Nk)^{1/2} + k^{1/2})$ per traced ray. Note that for most practical applications, $k \approx O(1)$ and $k \ll N$.

One of the advantages of the quantum solution is the fact that we do not require to use polygons. In classical rendering, polygons are used because it is easier to calculate intersections with them than with general objects. In the quantum solution, as long as we can establish that a certain object and a ray intersect in $O(1)$ time, we can use any type of objects. The quantum solution is then, much more robust for objects of completely arbitrary geometry. As in the case for Z-Buffering, one could argue that the use of a classical data structure can be used to speed up the classical algorithm. Table 4.3 shows a comparison between ray tracing algorithms, where N_r is the total number of rays that needs to be traced. Once more, the greatest advantage of the quantum solution resides on its optimal query time complexity for general objects with linear space complexity.

Table 4.3: Comparison of ray tracing algorithms. In this table, d is the dimension of the space, N is the number of polygons, k is the number of polygons intersected per ray traced, and N_r is the total number of traced rays.

Ray Tracing Algorithm	Query Time	Space Resources
Classical ray tracing for general objects	$O(N_r N)$	$O(N)$
Classical ray tracing using coordinate-aligned boxes and linear space trees	$O(N_r(N^{1-1/d} + k))$	$O(N)$
Classical ray tracing using coordinate-aligned boxes and non-linear space trees	$O(N_r(\log^d(N) + k))$	$O(N \log^{d-1} N)$
QMOS-based for general objects	$O(N_r((kN)^{1/2} + k^{1/2}))$	$O(N)$

4.2.3 RADIOSITY

As we mentioned before, ray tracing is an excellent method to simulate reflected and refracted light. However, a ray tracer can only approximate in a very crude and expensive way diffuse light (the light scattered by opaque materials). Radiosity is a method that models extremely well diffuse light but offers a poor representation of reflected light [12]. In other words, ray tracing is a good method to render images of shiny and semi- transparent materials, while radiosity is optimal to render opaque surfaces.

The radiosity method is based on the principle of energy conservation [12, 18]. In few words, radiosity is defined as the energy leaving each element in the scene. Radiosity embodies the idea that the total energy radiated by an opaque material is equal to the energy naturally emitted by the material plus the reflected energy:

Radiosity = emitted energy + reflected energy.

where:

Reflected energy = reflection coefficient * *total energy incident on the object from all other objects in the scene.*

This can be written in the form of the so-called *radiosity equation*, given by:

$$\mathbf{B}_i d\mathbf{A}_i \;=\; \mathbf{E}_i d\mathbf{A}_i \;+\; \zeta_i \int \mathbf{B}_j \mathbf{F}_{ji} d\mathbf{A}_j \;. \tag{4.6}$$

If we consider the scene to comprise a discrete number of individual objects (as we have already stated), the radiosity equation takes the form:

$$\mathbf{B}_i \mathbf{A}_i \;=\; \mathbf{E}_i \mathbf{A}_i \;+\; \zeta_i \Sigma_j \mathbf{B}_j \mathbf{F}_{ji} \mathbf{A}_j \;. \tag{4.7}$$

In these equations $\mathbf{B_i}$ is the radiosity of each element, $\mathbf{A_i}$ is the area of the element, $\mathbf{E_i}$ is the emitted energy and $\mathbf{F_{ij}}$ is the form factors matrix that determines the incident energy which was emitted by the other elements in the scene. The indices i and j are over all the elements that make the scene. The radiosity equation can also be rewritten in matrix form as:

$$\mathbf{M} \cdot \mathbf{B} \;=\; \mathbf{E} \tag{4.8}$$

where $\mathbf{M}$ is a form factors matrix. If we know $\mathbf{E}$ and $\mathbf{M}$, we can use the Gauss method to solve the system for the radiosity vector $\mathbf{B}$ in $O(N)$ steps. $\mathbf{E}$ is an intrinsic property of the material and, therefore, it is a known quantity. The form factors $\mathbf{M}$ are unknown but can be calculated by performing ray tracing between all the elements in the scene, which requires $O(N^2)$ steps in the classical case.

Now, once we know $\mathbf{B}$, we still need to calculate the shading of each pixel on the screen, which is accomplished by performing a technique very similar to ray tracing. So, the radiosity of a pixel P (which can be translated into a shading value) on the screen is given by:

$$Shading(P) \;=\; \mathbf{B} \cdot \mathbf{N}(P, xp) \tag{4.9}$$

where $\mathbf{N}(P, xp)$ is a vector that depends on the pixel P and the point of the scene that it intersects xp (Figure 4.8). Each element of $\mathbf{N}(xp)$ determines how much the radiosity of the element i contributes to the shading of the pixel P. So, for example, if an object is not visible from P, the value of $\mathbf{N}$ for that element is zero (it does not contribute to the shading of the pixel). To calculate the values of $\mathbf{N}(xp)$ we need to apply a ray tracer that determines the visibility of each object in the scene from pixel P, so it takes $O(N)$ operations to complete.

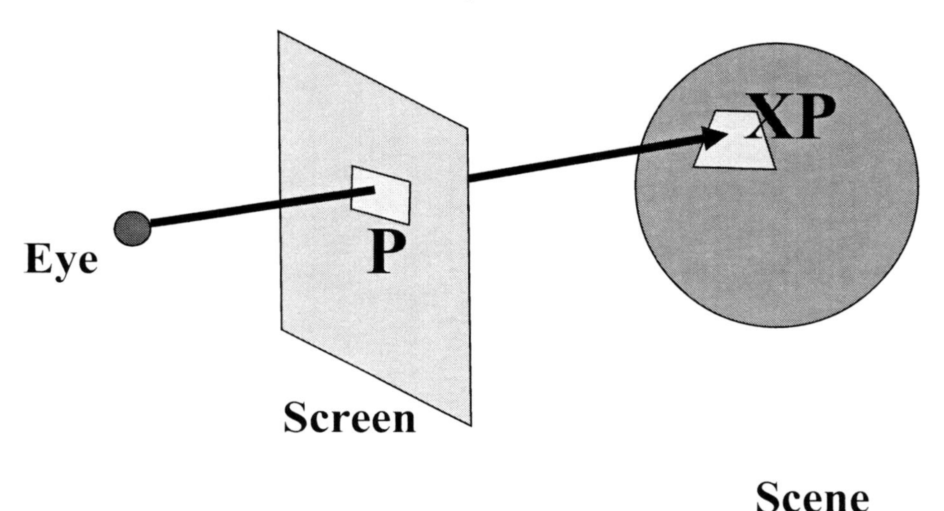

Figure 4.8: The shading of pixel P depends on the radiosities at xp.

Therefore, the shading of each pixel involves the following three operations:

1. An $O(N^2)$ operation to calculate the form factor matrix.
2. An $O(N)$ operation to determine $\mathbf{N}$.
3. An $O(N)$ operation to solve the radiosity matrix equation for $\mathbf{B}$.

As already discussed, (1) and (2) are operations similar to ray tracing, which we already know how to implement on a quantum computer. So, using a quantum ray tracer we can perform task (1) in $O(N^{3/2})$ steps and task (2) in $O\left(N^{1/2}\right)$ steps. However, we cannot use QC to further optimize task (3). Consequently, quantum radiosity can be performed in $O(N^{3/2})$ steps, in contrast to the $O(N^2)$ time required in the classical case. Even if the radiosity method is performed using multi-dimensional data structures, we can perform a similar analysis as the one done for ray tracing to determine the advantages of a quantum implementation.

The speed up achieved by using a quantum radiosity algorithm can be used to increase the realism of the rendered images. As discussed before, radiosity does not model accurately reflected lights. However, there is an extension to the radiosity method, known as *ray-traced radiosity* that correctly simulates reflections and refractions. As the name indicates, ray-traced radiosity implements both techniques into a single, more complex method. Ray traced radiosity is a *two pass* method. On the first pass we perform radiosity as usual, and on the second pass we take reflections into account. On the second pass, for each pixel, we shoot several rays and obtain the radiosity of the reflected rays (Figure 4.9). If we use M rays per pixel, the second pass requires $O(M * N * S)$ operations if we consider $S - 1$ reflections. Evidently, this is an extremely time consuming algorithm. However, a quantum ray traced radiosity algorithm is straightforward to implement, and we could obtain up to a square root speed up. The second pass is then performed using $O(M * S * N^{1/2})$ steps.

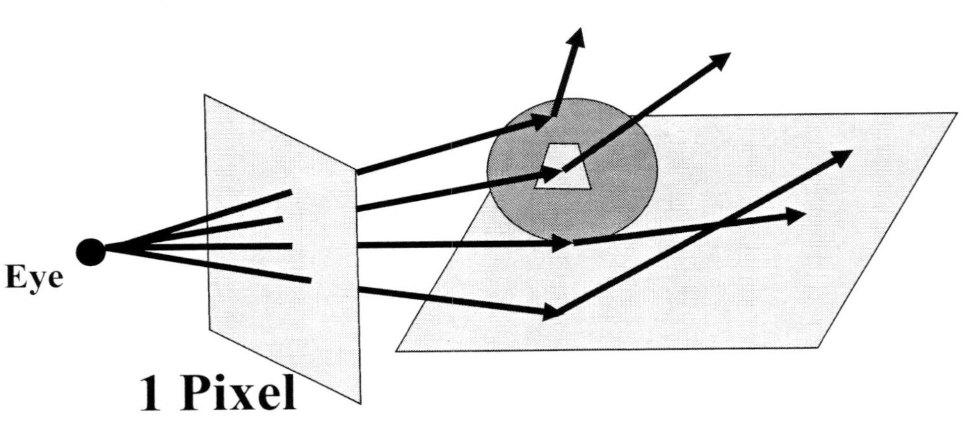

Figure 4.9: Ray traced radiosity for one pixel.

4.2.4 LEVEL OF DETAIL

Level of Detail (LOD) is not a rendering algorithm, but a scene management method [37]. LOD is based on the idea that, in many applications, we can trade fidelity for speed. For example, some details of an object may not be visible from the viewer's point of view, or these details may be so small that are imperceptible. At other times, we may purposely need a low-resolution image of the object to perform a quick visual exploration. In these cases, we can use a LOD algorithm that determines when and for how much we can simplify a given model in the scene.

If a polygonal mesh is used to describe the models, the LOD method determines what vertices in the mesh are necessary to represent the model according to the given circumstances. If some vertices are not required, the LOD eliminates them from the mesh. To determine if a vertex is necessary or not, the LOD function evaluates an error function $\epsilon(\nu)$ for each vertex ν. If the error associated with a vertex is less than a certain threshold $\delta < \epsilon$, then the vertex can be removed. This procedure has to be repeated for all the vertices in the mesh, for each frame being rendered. For increased performance, all the polygons are encoded in a hierarchical tree structure that determines which vertices are active at a given time. In this case, if the error is smaller than the threshold, the vertex may be expanded to include lower portions of the hierarchical tree. Nevertheless, the process has to be performed over the entire set of active vertices and requires $O(N)$ steps, where N is the number of active vertices.

At this point we can envision the quantum version of a LOD method. For each frame, we prepare a state $|\Psi>$ that encodes the entire mesh made of N vertices. We can use Hadamard gates to obtain a uniform superposition of the state. Each element of the superposition represents a vertex in the mesh. Now, we define a function f such that:

$$f(\nu) = 1 \text{ if } \epsilon(\nu) < \delta \tag{4.10}$$
$$f(\nu) = 0 \text{ any other case }. \tag{4.11}$$

We add some ancillary qubits and use quantum parallelism to evaluate $f(\nu)$ for each element of the superposition in a single step. Then, it becomes straightforward to apply quantum amplitude amplification algorithms to search for an unknown number of vertices with an error small enough that makes them removable. This operation can be performed with $O\left(N^{1/2}\right)$ steps, in contrast to the $O(N)$ steps required in the classical case. Of course, this procedure is used for each frame being rendered and N is the number of active vertices in the hierarchical tree structure.

4.3 SUMMARY

The QMOS algorithm provides the foundation for a general database system capable of satisfying *any* query with solution size k in $O((kN)^{1/2})$ time. As is well known, classical databases can only provide sublinear query complexity for predetermined classes of queries.

In particular, we discussed the applications of QMOS to computer graphics and presented a few quantum rendering algorithms. However, the applications of QMOS are varied and not limited to graphics. For instance, we have also developed quantum algorithms for target tracking in sonar applications, where it is not possible to structure the data in the form of coordinate-aligned boxes [58]. In this case, the best known classical method is not much better than exhaustive search, and a quantum solution offers a quadratic speed-up.

CHAPTER 5

The Quantum Fourier Transform

The Fourier Transform (FT) is regarded as among the most important tools for mathematicians, physicists and engineers. Indeed, this mathematical tool is widely used in virtually every single area of physics and engineering, including optics, high energy physics, acoustics, signal analysis, and image processing. Arguably, the ability to efficiently perform Fourier Transforms could radically change the amount of computational resources needed to solve a variety of problems.

Interestingly, the quantum computational model offers the possibility to perform Fourier Transforms exponentially faster than the best known classical method known today. However, it is important to note that this type of exponential speedup, so far appears to be practical only to algorithms that solve the Hidden Subgroup Problem (HSP). Therefore, it is quite unfortunate that the Quantum Fourier Transform cannot be directly applied in the areas mentioned above.

As we will see in the next chapter, we can use the quantum version of the Fourier Transform to efficiently solve the Hidden Subgroup Problem, which has important applications in cryptoanalysis. Indeed, by solving the Hidden Subgroup Problem exponentially faster than with the best known classical methods, the quantum computating model is able to compromise the most widely used asymmetric cryptosystems. This is, by far, the most important feature of the quantum computational model.

5.1 THE CLASSICAL FOURIER TRANSFORM

Before tackling the issue of Quantum Fourier Transforms, let us briefly recall the classical version of this important mathematical tool. To this end, suppose we have a set of N data points:

$$(x_0, x_1, ..., x_{N-1}) \ . \tag{5.1}$$

Then, the *discrete Fourier Transform* of the set of points is defined as:

$$(x_0, x_1, ..., x_{N-1}) \rightarrow (y_0, y_1, ..., y_{N-1}) \tag{5.2}$$

where the transformed points are described by:

$$y_k = \frac{1}{\sqrt{N}} \sum_{j=0}^{N-1} e^{i2\pi jk/N} x_j \ . \tag{5.3}$$

We can easily extend this concept to the realm of continue functions. That is, if instead of a set of discrete points we have a function $f(x)$, we can discretize it over N values as:

$$f(x) \rightarrow (f(0), f(1), ..., f(N-1)) \equiv f_N \,. \tag{5.4}$$

In this way, the Fourier Transform of $f(x)$ is defined by:

$$f_N \rightarrow \left(\hat{f}(0), \hat{f}(1), ..., \hat{f}(N-1)\right) \equiv \hat{f}_N \tag{5.5}$$

$$\hat{f}(y) = \frac{1}{\sqrt{N}} \sum_{x=0}^{N-1} e^{i2\pi xy/N} f(x) \,. \tag{5.6}$$

Therefore, the discrete Fourier Transform can be used to approximate the continuous Fourier Transform based on the N regular samples.

Clearly, the explicit calculation of $\hat{f}_N$ requires N operations for each of the N sample points. This implies that the overall complexity for the determination of the classical Fourier Transform is $FT(f_N) = O(N^2)$.

This quadratic complexity is prohibitive for large values of N. If we use $O(n) = O(Log(N))$ bits to represent the values of N, then the computational complexity to compute the Fourier Transform in function of n is $O(2^{poly(n)})$. That is, the number of computational steps grows exponentially with the number of bits necessary to represent the values of N.

However, this result can be improved using the *Fast Fourier Transform (FFT)*, which is often considered as the most important algorithmic development of the 20th century [11]. The FFT achieves better performance: $FFT(f_N) = O(N \log(N))$. However, when we consider the complexity of the computation in terms of the number of bits necessary to express the value of the N points, the complexity remains exponential.

However, as we will see with some detail in this chapter, it turns out that a quantum analog to the Fourier Transform can achieve even better complexity in certain special applications.

5.2 THE QUANTUM FOURIER TRANSFORM

The quantum version of the discrete Fourier Transform is constructed following the same rational as before [40, 5]. The first step is to find the Fourier Transform of the computational basis that spans a quantum superposition. Thus, let us suppose that we are using n-qubits representing $N = 2^n$ states, and we take the computational basis conveniently labeled as:

$$|\mathbf{0}\rangle, |\mathbf{1}\rangle, ..., |\mathbf{N-1}\rangle \,. \tag{5.7}$$

Now, we define the Quantum Fourier Transform (QFT) of the elements of the computational basis in a similar way to the discrete Fourier Transform:

$$QFT|\boldsymbol{k}\rangle = \frac{1}{\sqrt{N}} \sum_{j=0}^{N-1} e^{i2\pi jk/N} |\boldsymbol{j}\rangle \,. \tag{5.8}$$

Then, for a general quantum superposition the QFT looks like:

$$\sum_{k=0}^{N-1} x_k |\boldsymbol{k}\rangle \rightarrow \sum_{j=0}^{N-1} y_j |\boldsymbol{j}\rangle \tag{5.9}$$

where:

$$y_j = \frac{1}{\sqrt{N}} \sum_{r=0}^{N-1} e^{i2\pi rj/N} x_r \,. \tag{5.10}$$

Indeed:

$$\begin{aligned} QFT\left(\sum_{r=0}^{N-1} x_r |\boldsymbol{r}\rangle\right) &= \sum_{r=0}^{N-1} x_r QFT(|\boldsymbol{r}\rangle) & (5.11)\\ &= \sum_{r=0}^{N-1} x_r \left(\frac{1}{\sqrt{N}} \sum_{k=0}^{N-1} e^{i2\pi kr/N} |\boldsymbol{k}\rangle\right) & (5.12)\\ &= \sum_{k=0}^{N-1} \left(\frac{1}{\sqrt{N}} \sum_{r=0}^{N-1} e^{i2\pi rk/N} x_r\right) |\boldsymbol{k}\rangle & (5.13)\\ &= \sum_{k=0}^{N-1} y_k |\boldsymbol{k}\rangle \,. & (5.14) \end{aligned}$$

That is, the QFT of an arbitrary quantum state is defined as the superposition of the Fourier Transform of the individual amplitudes of the original state.

5.3 MATRIX REPRESENTATION

As we mentioned before, the matrix representation of quantum operators is very useful to compute their effect on arbitrary quantum states. In this section, we will obtain a matrix representation for the Quantum Fourier Transform and we will argument its unitarity.

To this end we can start with the previous definition of the Quantum Fourier Transform, and we expand the terms in the summation. It is clear that:

$$y_k = \frac{1}{\sqrt{N}} \left(e^{i2\pi k0/N} x_0 + e^{i2\pi k1/N} x_1 + \ldots + e^{i2\pi k(N-1)/N} x_{N-1}\right) \,. \tag{5.15}$$

From this equation is easy to obtain the matrix elements of the Quantum Fourier Transform, which are described by:

$$\begin{pmatrix} y_0 \\ y_1 \\ \ldots \\ y_{N-1} \end{pmatrix} = \frac{1}{\sqrt{N}} \begin{pmatrix} e^{i2\pi 00/N} & e^{i2\pi 01/N} & \ldots & e^{i2\pi 0(N-1)/N} \\ e^{i2\pi 10/N} & e^{i2\pi 11/N} & \ldots & e^{i2\pi 1(N-1)/N} \\ \ldots & \ldots & \ldots & \ldots \\ e^{i2\pi (N-1)0/N} & e^{i2\pi (N-1)1/N} & \ldots & e^{i2\pi (N-1)(N-1)/N} \end{pmatrix} \begin{pmatrix} x_0 \\ x_1 \\ \ldots \\ x_{N-1} \end{pmatrix} . \tag{5.16}$$

To obtain a more compact definition, it is useful to define:

$$\omega = e^{i2\pi/N} \tag{5.17}$$

because in this case we have that:

$$\begin{pmatrix} y_0 \\ y_1 \\ \dots \\ y_{N-1} \end{pmatrix} = \frac{1}{\sqrt{N}} \begin{pmatrix} \omega^{00} & \omega^{01} & \dots & \omega^{0(N-1)} \\ \omega^{10} & \omega^{11} & \dots & \omega^{0(N-1)} \\ \dots & \dots & \dots & \dots \\ \omega^{(N-1)0} & \omega^{(N-1)1} & \dots & \omega^{(N-1)(N-1)} \end{pmatrix} \begin{pmatrix} x_0 \\ x_1 \\ \dots \\ x_{N-1} \end{pmatrix} \tag{5.18}$$

and, therefore, the operator that describes the Quantum Fourier Transform can be simply written as:

$$QFT_{pq} = \frac{1}{\sqrt{N}} \left[\omega^{pq}\right] . \tag{5.19}$$

Using Plancherel's theorem it can be shown that the QFT is unitary as required[1].

5.4 CIRCUIT REPRESENTATION

In order to compute the computational complexity of the Quantum Fourier Transform, it is useful to rewrite the operator in terms of basic computing gates. To this end, we use the binary representation of an integer j:

$$j \equiv j_1 j_2 \dots j_n \equiv j_1 2^{n-1} + j_2 2^{n-2} + \dots + j_n 2^0 . \tag{5.20}$$

That is,

$$j_1 j_2 \dots j_n = \sum_{r=0}^{n-1} 2^r j_{n-r} . \tag{5.21}$$

Similarly, the binary fraction is given by:

$$0.j_l j_{l+1} \dots j_m \equiv \frac{j_l}{2} + \frac{j_{l+1}}{2^2} + \dots + \frac{j_m}{2^{m-l+1}} \tag{5.22}$$

which clearly satisfies:

$$0.j_1 j_2 \dots j_n = \frac{j_1}{2} + \frac{j_2}{2^2} + \dots + \frac{j_n}{2^n} \tag{5.23}$$

$$= \sum_{r=1}^{n} \frac{j_r}{2^r} \tag{5.24}$$

[1] In classical Fourier analysis, Plancherel's theorem states that, if a function f is in both $L^1(R)$ and $L^2(R)$, then its Fourier Transform is in $L^2(R)$, and the Fourier Transform mapping is isometric (preserves distances) and can be shown to be unitary. Here, $L^p(R)$ is the space of *p-power integrable functions*. However, the full proof that the classical Fourier Transform is an unitary transformation is usually known as Parseval's Theorem.

and therefore:

$$j_1 j_2 ... j_n = 2^n \, 0.j_1 j_2 ... j_n \ . \tag{5.25}$$

We can write any arbitrary state of the quantum basis using the binary representation:

$$|\boldsymbol{r}\rangle = |\boldsymbol{r_1 r_2 ... r_n}\rangle = |\boldsymbol{r_1 2^{n-1} + r_2 2^{n-2} + ... + r_n 2^0}\rangle \ . \tag{5.26}$$

Now, let us apply a Quantum Fourier Transform to an arbitrary state $|\boldsymbol{j}\rangle$ written in the binary representation:

$$\begin{aligned}
QFT|\boldsymbol{j}\rangle &= QFT|\boldsymbol{j_1 j_2 ... j_n}\rangle && (5.27)\\
&= \frac{1}{\sqrt{N}} \sum_{k=0}^{N-1} e^{i2\pi kj/N} |\boldsymbol{k}\rangle && (5.28)\\
&= \frac{1}{2^{n/2}} \sum_{k=0}^{2^n-1} e^{i2\pi kj/2^n} |\boldsymbol{k}\rangle && (5.29)\\
&= \frac{1}{2^{n/2}} \sum_{k=0}^{2^n-1} e^{i2\pi j \sum_{r=0}^{n-1} k_r/2^r} |\boldsymbol{k}\rangle && (5.30)\\
&= \frac{1}{2^{n/2}} \sum_{k=0}^{2^n-1} \prod_{r=1}^{n} e^{i2\pi j k_r/2^r} |\boldsymbol{k}\rangle && (5.31)\\
&= \frac{1}{2^{n/2}} \sum_{k_1=0}^{1} ... \sum_{k_n=0}^{1} \prod_{r=1}^{n} e^{i2\pi j k_r/2^r} |\boldsymbol{k_1 ... k_n}\rangle && (5.32)\\
&= \frac{1}{2^{n/2}} \sum_{k_1=0}^{1} ... \sum_{k_n=0}^{1} \prod_{r=1}^{n} e^{i2\pi j k_r/2^r} |\boldsymbol{k_r}\rangle && (5.33)\\
&= \frac{1}{2^{n/2}} \prod_{r=1}^{n} \sum_{k_r=0}^{1} e^{i2\pi j k_r/2^r} |\boldsymbol{k_r}\rangle && (5.34)\\
&= \frac{1}{2^{n/2}} \prod_{r=1}^{n} \left(|\boldsymbol{0}\rangle + e^{i2\pi j/2^r} |\boldsymbol{1}\rangle\right) && (5.35)\\
&= \frac{1}{2^{n/2}} \left(|\boldsymbol{0}\rangle + e^{i2\pi j/2} |\boldsymbol{1}\rangle\right) ... \left(|\boldsymbol{0}\rangle + e^{i2\pi j/2^n} |\boldsymbol{1}\rangle\right) && (5.36)\\
&= \frac{1}{2^{n/2}} \left(|\boldsymbol{0}\rangle + e^{i2\pi 0.j_n} |\boldsymbol{1}\rangle\right) ... \left(|\boldsymbol{0}\rangle + e^{i2\pi 0.j_1 ... j_n} |\boldsymbol{1}\rangle\right) && (5.37)
\end{aligned}$$

where the last equation is true because, for the last term:

$$\frac{j}{2^n} = \frac{j_1 ... j_n}{2^n} = 0.j_1 ... j_n \ . \tag{5.38}$$

And for the previous term:

$$\frac{j}{2^{n-1}} = 2\frac{j_1 \dots j_n}{2^n} \tag{5.39}$$
$$= 2\left(\frac{j_1}{2} + \frac{j_2}{2^2} + \dots + \frac{j_n}{2^n}\right) \tag{5.40}$$
$$= j_1 + \frac{j_2}{2} + \dots + \frac{j_n}{2^{n-1}} \tag{5.41}$$
$$= j_1 + 0.j_2 \dots j_n \ . \tag{5.42}$$

And therefore:

$$e^{i2\pi j/2^{n-1}} = e^{i2\pi(j_1+0.j_2\dots j_n)} \tag{5.43}$$
$$= e^{i2\pi j_1} e^{i2\pi 0.j_2\dots j_n} \tag{5.44}$$
$$= e^{i2\pi 0.j_2\dots j_n} \ . \tag{5.45}$$

And similarly for all the other terms. Therefore, we can write the Quantum Fourier Transform of the basis state in the *Product Representation* as:

$$|\boldsymbol{j_1 j_2 \dots j_n}\rangle \xrightarrow{QFT} \frac{\left(|\mathbf{0}\rangle + 2^{i2\pi 0.j_n}|\mathbf{1}\rangle\right)\left(|\mathbf{0}\rangle + 2^{i2\pi 0.j_{n-1}j_n}|\mathbf{1}\rangle\right)\dots\left(|\mathbf{0}\rangle + 2^{i2\pi 0.j_1 j_2\dots j_n}|\mathbf{1}\rangle\right)}{2^{n/2}} \ . \tag{5.46}$$

This representation is often considered a definition of the QFT. The product representation is useful because it can be used to define an efficient quantum circuit to compute QFTs. The most important unit of this circuit is the R_k gate:

$$R_k \equiv \begin{pmatrix} 1 & 0 \\ 0 & e^{i2\pi/2^k} \end{pmatrix} . \tag{5.47}$$

The quantum circuit that represents the QFT is depicted in Figure 5.1, and it involves Hadamard and R_k gates. This circuit needs as an input the state $|\boldsymbol{j_1 \dots j_n}\rangle$. By using the quantum circuit analysis technique we described on a previous chapter, it can be determined that the output state is:

$$\frac{\left(|\mathbf{0}\rangle + 2^{i2\pi 0.j_1 j_2\dots j_n}|\mathbf{1}\rangle\right)\dots\left(|\mathbf{0}\rangle + 2^{i2\pi 0.j_{n-1}j_n}|\mathbf{1}\rangle\right)\left(|\mathbf{0}\rangle + 2^{i2\pi 0.j_n}|\mathbf{1}\rangle\right)}{2^{n/2}} \ . \tag{5.48}$$

Clearly, this is the QFT of $|\boldsymbol{j_1 \dots j_n}\rangle$, but the order of the qubits has been swaped. Therefore, to obtain the QFT we only require to perform a swaping on all the qubits. The circuit that swaps two qubits is made of three concatenated CNOT as depicted in Figure 5.2.

Indeed, if the initial state is:

$$|\boldsymbol{\Psi_0}\rangle = |\boldsymbol{\Psi_a}\rangle \otimes |\boldsymbol{\Psi_b}\rangle \tag{5.49}$$
$$= (\alpha|\mathbf{0}\rangle + \beta|\mathbf{1}\rangle) \otimes (\gamma|\mathbf{0}\rangle + \delta|\mathbf{1}\rangle)) \tag{5.50}$$
$$= (\alpha\gamma|\mathbf{00}\rangle + \alpha\delta|\mathbf{01}\rangle + \beta\gamma|\mathbf{10}\rangle + \beta\delta|\mathbf{11}\rangle) \ . \tag{5.51}$$

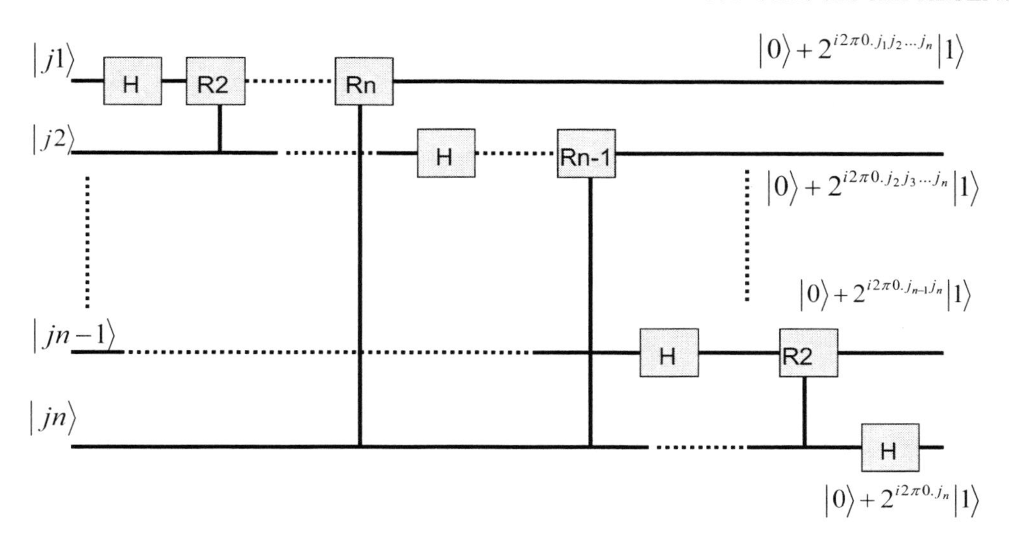

Figure 5.1: Diagram of the quantum circuit that implements a Quantum Fourier Transform on the state $|\boldsymbol{j_1 \ldots j_n}\rangle$.

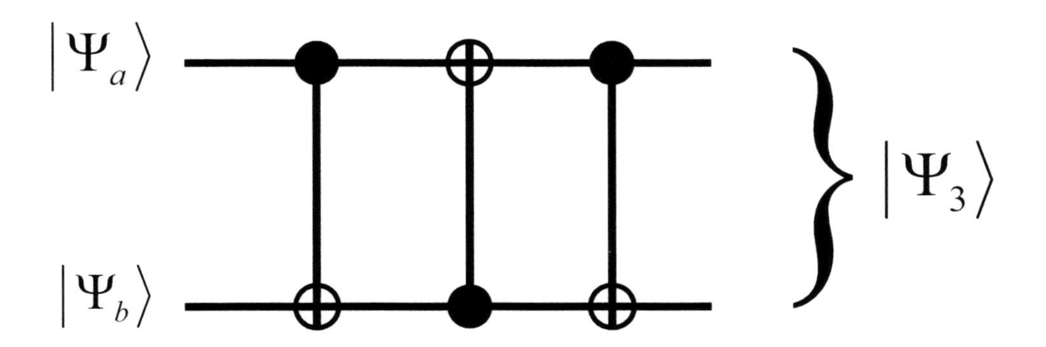

Figure 5.2: Circuit that swaps the order of two qubits.

After the first CNOT gate:

$$|\boldsymbol{\Psi_1}\rangle = (\alpha\gamma|\mathbf{00}\rangle + \alpha\delta|\mathbf{01}\rangle + \beta\gamma|\mathbf{11}\rangle + \beta\delta|\mathbf{10}\rangle) \tag{5.52}$$

and after the second CNOT gate:

$$|\boldsymbol{\Psi_2}\rangle = (\alpha\gamma|\mathbf{00}\rangle + \alpha\delta|\mathbf{11}\rangle + \beta\gamma|\mathbf{01}\rangle + \beta\delta|\mathbf{10}\rangle) \tag{5.53}$$

and after the third and final CNOT gate:

$$\begin{aligned} |\Psi_3\rangle &= (\alpha\gamma|\mathbf{00}\rangle + \alpha\delta|\mathbf{10}\rangle + \beta\gamma|\mathbf{01}\rangle + \beta\delta|\mathbf{11}\rangle) && (5.54)\\ &= (\gamma|\mathbf{0}\rangle + \delta|\mathbf{1}\rangle)) \otimes (\alpha|\mathbf{0}\rangle + \beta|\mathbf{1}\rangle) && (5.55)\\ &= |\Psi_b\rangle \otimes |\Psi_a\rangle && (5.56) \end{aligned}$$

and as intended, the qubits have been swaped.

5.5 COMPUTATIONAL COMPLEXITY

In order to have an objective comparison of the algorithmic advantages of the Quantum Fourier Transform over its classical counterpart, we need to determine its circuit and algorithmic complexity.

To determine the circuit complexity, we need to consider the circuit that represents the Quantum Fourier Transform discussed on the previous section. We can do a counting of the number of gates that are necessary to implement this circuit:

- For the 1st qubit we need 1 Hadammard gate and $(n-1)$ R_k gates.
- For the 2nd qubit we need 1 Hadammard gate and $(n-2)$ R_k gates.
- ...
- For the nth qubit we need 1 Hadammard gate and 0 R_k gates.

Thus, we need n Hadamard gates and:

$$(n-1) + (n-2) + \ldots + 1 = \frac{n(n+1)}{2} \tag{5.57}$$

R_k gates. In addition, we require to swap the qubits, as the transformation is in inverse order (from the lowest qubit to the higher qubit). It can easily be shown that we require about $3n/2$ swap gates to accomplish this. Therefore, the total number of gates necessary to represent a circuit that performs a Quantum Fourier Transform is:

$$\frac{3n}{2} + \frac{n(n+1)}{2} \approx O(n^2) \,. \tag{5.58}$$

Therefore, this circuit has a gate complexity of $O(n^2) = O(\log^2(N))$. If each gate takes a single computational step, then the quantum algorithm that performs the QFT requires $O(\log^2(N))$ computational steps. Obviously, this $O(\log^2(N))$ complexity represents an exponential speedup over the $O(N\log(N))$ complexity of the classical FFT algorithm.

5.6 ALGORITHMIC RESTRICTIONS

On the previous section we showed that the Quantum Fourier Transform can be computed exponentially faster than its classical counterpart. Unfortunately, the QFT suffers a variety of constraints which severely limit its practical utility.

5.6.1 NORMALIZATION

The first constraint is that the Quantum Fourier Transform of the set of points:

$$(x_0, x_1, ..., x_{N-1}) \rightarrow (y_0, y_1, ..., y_{N-1}) \tag{5.59}$$

requires the numbers to have amplitudes normalized to unity:

$$x_0^2 + x_1^2 + ... + x_{N-1}^2 = 1 \tag{5.60}$$

$$y_0^2 + y_1^2 + ... + y_{N-1}^2 = 1\ . \tag{5.61}$$

If you think about it, this is a rather severe restriction. Clearly, for most practical applications, the data will not be normalized to the unity.

However, it is important to note that one could easily renormalize the input data to make it unitary:

$$(x_0, x_1, ..., x_{N-1}) \rightarrow (x_0^{(n)}, x_1^{(n)}, ..., x_{N-1}^{(n)}) \tag{5.62}$$

$$= \frac{(x_0, x_1, ..., x_{N-1})}{norm(X)} \tag{5.63}$$

$$= \frac{(x_0, x_1, ..., x_{N-1})}{\sqrt{x_0^2 + x_1^2 + ... + x_{N-1}^2}}\ . \tag{5.64}$$

Then, by Parseval Theorem, we know that the Fourier Transform will also be unitary:

$$y_0^2 + y_1^2 + ... + y_{N-1}^2 = 1\ . \tag{5.65}$$

However, the normalization step, which involves the explicit calculation and addition of the magnitudes of each of the N elements, is a process that has to be computed with $\Theta(N)$ computational steps. Thus, the normalization of the data overrides the benefit of using a Quantum Fourier Transform.

5.6.2 INITIALIZATION

The second constraint is that, because of Property # 8, a quantum register can only be initialized to "0". Thus, we need to transform the "0" state into the desired initial state. To do this we require to determine the elements of a $N \times N$ unitary transformation by solving a linear system of N equations and N variables. This process has $O(N^2)$ complexity, which once more, completely negates the computational advantages of the QFT. Of course, depending on the application, one could pre-compute the initialization matrix once, and then use this representation on subsequent applications of the QFT in $O(1)$ time.

5.6.3 OUTPUT

The third constraint is that the QFT encodes the Fourier transformation in the amplitudes:

$$(x_0, x_1, ..., x_{N-1}) \rightarrow (y_0, y_1, ..., y_{N-1}) \tag{5.66}$$

$$\sum_{j=0}^{N-1} x_j |\boldsymbol{j}\rangle \rightarrow \sum_{k=0}^{N-1} y_k |\boldsymbol{k}\rangle \ . \tag{5.67}$$

And there is no efficient way to extract the information out of the amplitudes. Indeed, the value of the amplitude is merely related to the probability of observing that state upon measurement of the quantum register. Because a measurement results in the collapse of the superposition and we cannot make copies of the register, there is no efficient way to extract all the information from the quantum register.

Of course, one could try other ways to extract this information. For instance, we could try to repeat the application of the Quantum Fourier Transform several times, and after performing measurements, we could calculate the probabilities. However, to do this we would have to repeat the $O(Log^2(N))$ QFT process for at least $O(N Log(N))$ times, for an overall complexity of $O(N Log^3(N))$, which is suboptimal to the use of a classical FFT.

5.7 SUMMARY

In summary, the QFT offers an exponential speedup over the classical FFT. Unfortunately, the QFT cannot be used to replace the FFT in general applications in the areas of optics, acoustics, and signal analysis. Indeed, the QFTs are useful only when the problem at hand does not require the list of all the elements of the Fourier transform. That is, it is very subtle how QFTs provide algorithmic advantages over standard classical methods.

Fortunately, despite these constraints, there are a few practical applications of the QFT, which are potentially very important. Specifically, the QFT can be used to find the period of a function, which in turn can be used to find the prime factorization of a large integer. Therefore, as we will see in the next chapter, QFTs can be used to break the most commonly used crytographic ciphers.

CHAPTER 6

Case Study: The Hidden Subgroup

As we discussed on the previous chapter, it is not feasible to apply Quantum Fourier Transforms to solve problems in optics, signal analysis, or electrodynamics. Indeed, the Quantum Fourier Transform suffers from unsurmountable deficiencies that prevent it from being used to accelerate classical applications in a trivial manner.

If you think about, the Quantum Fourier Transform is pretty much restricted to the types of problems that do not involve a pre-selected initial set of data, and which do not require as an output the actual values of the transformation. Even though these are limiting factors, there is an important application of this quantum computational technique.

Indeed, the Quantum Fourier Transform (QFT) can solve the *Hidden Subgroup Problem* (HSP) exponentially faster than is possible with the best available classical algorithm. While the Hidden Subgroup Problem does not appear too frequently in scientific or engineering problems, it is the backbone of the most widely used cryptosystems. That is, the Quantum Fourier Transform can be used to efficiently break our current standards of secure communications.

Before stating the formal expression of the Hidden Subgroup Problem, we will examine two illustrative problems: phase estimation and period finding.

6.1 PHASE ESTIMATION

Suppose we have an unitary operator U and a quantum state $|\boldsymbol{u}\rangle$ such that:

$$U|\boldsymbol{u}\rangle = e^{i2\pi\phi}|\boldsymbol{u}\rangle \tag{6.1}$$

and the goal is to estimate the value of the phase ϕ, where $0 \leq \phi \leq 1$. Clearly, this problem is somewhat similar to finding the eigenvalues of U, but in this case the eigenvalues are complex numbers restricted to have the form $\lambda = e^{i2\pi\phi}$.

To construct an efficient quantum algorithm that solves this problem we will require, as part of the input, a black box function and some ancillary qubits.

Let us assume that the black-box function performs a controlled U^j operation. That is, the operator U to the jth power. Thus:

$$|\boldsymbol{j}\rangle|\boldsymbol{u}\rangle \rightarrow |\boldsymbol{j}\rangle U^j|\boldsymbol{u}\rangle \ . \tag{6.2}$$

Let us also suppose that we have a 2^n-dimensional eigenstate u such that:

$$U|\boldsymbol{u}\rangle = e^{i2\pi\phi}|\boldsymbol{u}\rangle \ . \tag{6.3}$$

Now, we also assume that we have access to t extra qubits. Where the value of t is determined by the following relation:

$$t \approx n + \log\left(2 + \frac{1}{2\varepsilon}\right) . \tag{6.4}$$

The meaning of the value of ϵ will be evident later.

For the quantum algorithm to solve the phase estimation problem, we begin with an initial state which is made of the tensor product between the t ancillary qubits in the "0" state and the n-qubit target state $|\boldsymbol{u}\rangle$:

$$|\boldsymbol{\Psi_1}\rangle = |\boldsymbol{0}\rangle|\boldsymbol{u}\rangle \tag{6.5}$$

and then we create a uniform superposition over the t extra qubits using t Hadamard gates:

$$\begin{aligned} |\boldsymbol{\Psi_1}\rangle \rightarrow |\boldsymbol{\Psi_2}\rangle &= H^{\otimes t}|\boldsymbol{\Psi_1}\rangle & (6.6)\\ &= \left(H^{\otimes t}|\boldsymbol{0}\rangle\right)|\boldsymbol{u}\rangle & (6.7)\\ &= \frac{1}{\sqrt{2^t}}\sum_{j=0}^{2^t-1}|\boldsymbol{j}\rangle|\boldsymbol{u}\rangle . & (6.8) \end{aligned}$$

We then apply the black box function with the controlled U^j operator to obtain:

$$\begin{aligned} |\boldsymbol{\Psi_2}\rangle \rightarrow |\boldsymbol{\Psi_3}\rangle &= \frac{1}{\sqrt{2^t}}\sum_{j=0}^{2^t-1}|\boldsymbol{j}\rangle U^j|\boldsymbol{u}\rangle & (6.9)\\ &= \frac{1}{\sqrt{2^t}}\sum_{j=0}^{2^t-1}\left(e^{i2\pi\phi}\right)^j|\boldsymbol{j}\rangle|\boldsymbol{u}\rangle & (6.10)\\ &= \frac{1}{\sqrt{2^t}}\sum_{j=0}^{2^t-1}e^{i2\pi(j\phi)}|\boldsymbol{j}\rangle|\boldsymbol{u}\rangle . & (6.11) \end{aligned}$$

Now, let us suppose for a moment that the phase ϕ is a number that can be represented exactly with t bits on a binary fraction representation such as:

$$\phi = 0.\phi_1\phi_2...\phi_t . \tag{6.12}$$

In this case, it is easy to check that we can change to the standard binary representation by using a multiplicative factor 2^t:

$$0.\phi_1\phi_2...\phi_t = \frac{1}{2^t}\phi_1\phi_2...\phi_t . \tag{6.13}$$

Thus, we can define the shifted phase ϕ_c as:

$$\phi_c \equiv \phi_1\phi_2...\phi_t = 2^t \times 0.\phi_1\phi_2...\phi_t . \tag{6.14}$$

Then, using the standard definition of the Quantum Fourier Transform for ϕ_c we obtain:

$$|\boldsymbol{\phi_c}\rangle = |\boldsymbol{\phi_1 \phi_2 ... \phi_t}\rangle \xrightarrow{QFT} \frac{1}{\sqrt{2^t}} \sum_{j=0}^{2^t-1} e^{i2\pi(j\phi_c)/2^t} |\boldsymbol{j}\rangle \ . \tag{6.15}$$

Therefore, using this result we can see that the state $|\boldsymbol{\Psi_3}\rangle$ of our quantum algorithm takes a very convenient form:

$$|\boldsymbol{\Psi_3}\rangle = QFT\,(|\boldsymbol{\phi_1 \phi_2 ... \phi_t}\rangle) \otimes |\boldsymbol{u}\rangle \ . \tag{6.16}$$

If we now apply the inverse Quantum Fourier Transform we get:

$$|\boldsymbol{\Psi_4}\rangle = QFT^{-1}|\boldsymbol{\Psi_3}\rangle = |\boldsymbol{\phi_c}\rangle|\boldsymbol{u}\rangle \ . \tag{6.17}$$

Thus, the final step is to perform a measurement. However, notice how the state is no longer on a superposition, so the result of the measurement will be:

$$|\boldsymbol{\phi_c}\rangle|\boldsymbol{u}\rangle \tag{6.18}$$

with probability equal to 1. By measuring these states, we can obviously determine the value of ϕ_c, and the phase of the problem is determined by a simple arithmetic operation:

$$\phi = \frac{\phi_c}{2^t} \ . \tag{6.19}$$

Therefore, if the phase ϕ can be expressed exactly with t qubits, this quantum algorithm outputs the exact value of ϕ with probability one.

Clearly, the overall complexity of this quantum algorithm is dominated by the application of the inverse QFT, which has $O(t^2) = O(n^2)$ complexity. On the other hand, a classical algorithm to solve this same problem would have required $O(2^{poly(n)})$ time complexity. Thus, the quantum solution is exponentially faster.

Now, if the phase cannot be expressed exactly with t qubits, then it can be shown that the output after $O(t^2)$ computational steps is just an n-bit approximation to the value of ϕ

$$\phi_{\text{OUT}} \approx \phi \tag{6.20}$$

and the output is correct with probability

$$P = 1 - \varepsilon \ . \tag{6.21}$$

The definition of t in function of n and ε should be clear now. That is, the higher the precision we require for the output of the quantum algorithm, the more ancillary qubits we are required to have.

6.2 PERIOD FINDING

Suppose a binary function f is periodic over a finite domain:

$$f : \{0, 1\}^n \rightarrow \{0, 1\} \tag{6.22}$$

$$f(x) = f(x + r) \tag{6.23}$$

$$0 < r < 2^n . \tag{6.24}$$

As its name suggests, the period finding problem requires the determination of the value of the period r.

The quantum algorithm that we will use to solve the period finding problem is actually strikingly similar to the one used to solve the phase estimation problem.

As part of our input we will request t extra qubits, where t is given by:

$$t = O(n + \log(1/\varepsilon)) . \tag{6.25}$$

Just as before, the value of ε will be related to the precision of the algorithm. We also request a black-box function implemented on an unitary operator U that computes the value of f in the following fashion:

$$U|\boldsymbol{x}\rangle|\boldsymbol{y}\rangle = |\boldsymbol{x}\rangle|\boldsymbol{y} \oplus \boldsymbol{f}(\boldsymbol{x})\rangle \tag{6.26}$$

where $|\boldsymbol{x}\rangle$ is a quantum register of size t, and $|\boldsymbol{y}\rangle$ is a 1-qubit register that holds the output of the single-bit value of f.

As usual, the quantum algorithm begins with an initial state in the "0" position:

$$|\boldsymbol{\Psi}\rangle = |\boldsymbol{0}\rangle_{t\text{-qubits}}|\boldsymbol{0}\rangle_{1\text{-qubit}} . \tag{6.27}$$

For clarity, the subindex indicates the number of qubits in the specified register. Then, we create a uniform superposition on the t-register using t Hadamard gates:

$$|\boldsymbol{\Psi_2}\rangle = H^{\otimes t}|\boldsymbol{0}\rangle_t|\boldsymbol{0}\rangle_1 = \frac{1}{\sqrt{2^t}}\sum_{x=0}^{2^t-1}|\boldsymbol{x}\rangle_t|\boldsymbol{0}\rangle_1 . \tag{6.28}$$

And we, then, apply the U operator:

$$|\boldsymbol{\Psi_3}\rangle = U|\boldsymbol{\Psi_2}\rangle = \frac{1}{\sqrt{2^t}}\sum_{x=0}^{2^t-1}U|\boldsymbol{x}\rangle_t|\boldsymbol{0}\rangle_1 \tag{6.29}$$

$$= \frac{1}{\sqrt{2^t}}\sum_{x=0}^{2^t-1}|\boldsymbol{x}\rangle_t|\boldsymbol{f}(\boldsymbol{x})\rangle_1 . \tag{6.30}$$

We can approximate the value of $f(x)$ using its Fourier representation:

$$|\boldsymbol{f}(\boldsymbol{x})\rangle_1 \approx \frac{1}{\sqrt{r}} \sum_{l=0}^{r-1} e^{i2\pi(lx/r)} |\hat{\boldsymbol{f}}(\boldsymbol{l})\rangle_1 \tag{6.31}$$

and, therefore, the state that describes the quantum algorithm is given by:

$$|\boldsymbol{\Psi_3}\rangle \approx \frac{1}{\sqrt{r2^t}} \sum_{l=0}^{r-1} \sum_{x=0}^{2^t-1} e^{i2\pi(l/r)x} |\boldsymbol{x}\rangle_l |\hat{\boldsymbol{f}}(\boldsymbol{l})\rangle_1 \,. \tag{6.32}$$

Similarly, as before, if we can write the exact value of l/r using t bits, then we can conveniently define a shifted value of this ratio:

$$\left(\frac{l}{r}\right) = 0.\left(\frac{l}{r}\right)_1 \left(\frac{l}{r}\right)_2 \cdots \left(\frac{l}{r}\right)_t = \frac{1}{2^t}\left(\frac{l}{r}\right)_c \,. \tag{6.33}$$

And again, we observe that in this case, the Quantum Fourier Transform of the shifted ratio is given by:

$$QFT|\,(l/r)_c\rangle = \frac{1}{\sqrt{2^t}} \sum_{x=0}^{2^t-1} e^{i2\pi(l/r)_c x/2^t} |\boldsymbol{x}\rangle_t \tag{6.34}$$

which means that the quantum state that represents the state of the algorithm is given by:

$$|\boldsymbol{\Psi_3}\rangle \approx \frac{1}{\sqrt{r}} \sum_{l=0}^{r-1} QFT|\left(\tfrac{l}{r}\right)_c\rangle |\hat{\boldsymbol{f}}(\boldsymbol{l})\rangle \,. \tag{6.35}$$

Thus, applying the inverse Quantum Fourier Transform gives:

$$|\boldsymbol{\Psi_4}\rangle = QFT^{-1}|\boldsymbol{\Psi_3}\rangle \approx \frac{1}{\sqrt{r}} \sum_{l=0}^{r-1} |\left(\tfrac{l}{r}\right)_c\rangle_t |\hat{\boldsymbol{f}}(\boldsymbol{l})\rangle_1 \,. \tag{6.36}$$

Therefore, a measurement of this state will give the value of the ratio:

$$\left(\frac{l}{r}\right)_c \tag{6.37}$$

for some value of l.

At this point, we know an estimate to the value of l/r to a precision of t bits. However, we can also argue that l/r is a rational number. Indeed, l is the index that enumerates the Fourier expansion of f, and r is the period of f in the range $0 < r < 2^n$.

Then, to obtain the value of the period r, we need to apply the continued fractions algorithm, which outputs the nearest rational fraction to l/r [1]. This algorithm takes $O(t^3)$ computational

[1] A detailed explanation of this algorithm can be found in [40].

steps to complete. So, in this case, the complexity of the quantum algorithm is dominated by the classical part, which at $O(t^3)$, it overrides the complexity of applying the inverse Quantum Fourier Transform.

It is important to note that there is a fixed, small probability that the continued fractions step may fail, which will require us to repeat the entire algorithm. However, the expected number of iterations is $O(1)$ and so does not affect the overall complexity, which is dominated by the $O(t^3) = O(n^3)$ complexity of the continued fractions step. On the other hand, the best known classical algorithm that finds the period of a function takes $O(2^{poly(n)})$ time.

6.3 THE HIDDEN SUBGROUP PROBLEM

As we have just seen, the quantum algorithms that solve the phase estimation and period finding problems are very similar. This is because both problems are specific instances of a more general type of problem, often known as the *Hidden Subgroup Problem*.

The Hidden Subgroup Problem can be formally stated in the following way [1]. Suppose G is a group, K is a subgroup of G, and X is a finite set of elements, and f is a function such that:

$$f : G \rightarrow X \tag{6.38}$$
$$f(x) = f(y) \Leftrightarrow xK = yK \tag{6.39}$$
$$xK \equiv \{xk | k \in K\} \qquad yK \equiv \{yk | k \in K\} . \tag{6.40}$$

The solution to the Hidden Subgroup Problem means to find the generators for the group K.

Even though it is a very formal definition, it is easy to see how the problem of finding the period of a function f is a specific instance of the Hidden Subgroup Problem. Specifically, let us take the group G as the set of the positive natural integers with the standard addition as the group operation.

$$G = Z_N . \tag{6.41}$$

Now, let us consider K as a subgroup of G, such that the elements of K are the values of the period of a periodic function f. Such a periodicity has to be valid for all the domain values of f in G. That is:

$$K = \{k \in G | f(k + g) = f(g), \forall g \in G\} . \tag{6.42}$$

Therefore, finding the generators of K is equivalent to finding the period of the function f.

By choosing an specific group G and a subgroup K, and asking on each case to determine the generators of K, we can generate different problems [40]. Each of these problems is, therefore, an instance of the Hidden Subgroup Problem. Some other instances of the Hidden Subgroup Problem include:

- Phase Estimation Problem
- Period Finding Problem

- Order Finding Problem
- Discrete Logarithm Problem
- Order of a Permutation Problem
- Hidden Linear Function Problem
- Deutsch Problem
- Simon Problem

The quantum solution to any of these instances of the Hidden Subgroup Problem follows the same steps as we presented for the period finding and phase estimation problems [59]. That is, (1) we add extra qubits to the register, (2) establish the functionality of a black box function that represents the specific problem, (3) manipulate the state to find an expression that resembles a Quantum Fourier Transform, (4) apply the inverse Quantum Fourier Transform, (5) measure the register, and (6) perform any arithmetic operations necessary to obtain a meaningful result.

It can be formally shown that for the case when G is a finite Abelian group, a quantum computer can solve the Hidden Subgroup Problem in $O(\log |G|)$ time [1]. As it happens, this is exponentially faster than is possible for any known classical method.

It is important to note, however, that a few types of non-Abelian groups (e.g., the symmetric and dihedral groups) have been successfully solved with subexponential complexity, but it is not known whether the QFT can provide similar efficiency for other non-Abelian groups [39].

One could be easily misled to believe that the instances of the Hidden Subgroup Problem listed above are only of theoretical interest to the computer scientist, with little practical application. However, as we will see next, there is an important class of practical problems that involve instances of the Hidden Subgroup Problem. And as a matter of fact, this application of quantum algorithmics may well be the most critical result in the area of quantum computing.

6.4 QUANTUM CRYPTOANALYSIS

Because the Hidden Subgroup Problem is difficult to solve in the classical realm, several of its instances have been used in the past for cryptographical applications. Therefore, it is very important the result that a quantum computer can potentially solve the Hidden Subgroup Problem exponentially faster. In other words, the most critical practical application of quantum Hidden Subgroup Problem solvers is to break crypto systems that rely on the assumption that finding the solution to the Hidden Subgroup Problem requires exponential time.

However, it is important to note that although the quantum algorithms for the solution of the Hidden Subgroup Problem are more efficient than any known classical approaches, there is at present time no theoretical lower bound that prevents the derivation of a classical alternative with the same computational efficiency. That is, in the future someone could come up with a classical algorithm to solve the Hidden Subgroup Problem in logarithmic time.

In this section we will briefly discuss how the Hidden Subgroup Problem relates to cryptoanalysis. We recall that there are two major types of ciphers to encrypt secret information: symmetric and asymmetric. The symmetric ciphers have a single key, which is kept secret, and can be used for both, encrypting and decrypting secret messages. Some of these ciphers can be proved to be *perfectly secure*. That is, they are secure regardless of the computational power available to the adversaries. However, the problem with symmetric ciphers is the distribution of the secret key among the legit users. In most cases, a trusted courier has to deliver the secret keys, which according to the circumstances, it may be difficult, expensive, or unfeasible to accomplish.

On the other hand, asymmetric ciphers have two keys, a public key for encryption and a secret key for decryption. The encryption key is made public, so anybody can encrypt a secret message. An encrypted message is sent to the legal receiver, who uses his secret key to decrypt the message. Because the encryption key is public, its distribution is easy to accomplish. However, it is impossible to prove these asymmetric ciphers to be *perfectly secure*, because their security will always depend on assumptions regarding the amount of computational resources available to the adversary.

These asymmetric ciphers very often are designed in such a way that breaking them to obtain the secret key implies solving a hard mathematical problem. Let us consider, for example, the case of RSA, which probably is the most important and widely used cryptosystem. RSA and its variants form the current standards for public key distribution ciphers. Without going into details, the security of RSA is based on the computationally difficult problem of finding the prime factorization of a large coprime integer (an integer that results from the product of two prime numbers). The factorization problem is *believed* to be very difficult to solve using classical resources, and therefore, the security of RSA is based on an unproved mathematical assumption.

Indeed, if we use brute force to find the two prime factors, p and q, of a large coprime integer r represented with n-bits, we will require about $O(\sqrt{2^n}) = O(2^{n/2})$ computational steps (basically we need to try, one by one, all the integers from 1 to $\sqrt{r}$). Being exponential in the size of the integer n, the amount of computational resources necessary to break the cipher grows very rapidly. For instance, if the size of the key is 4 bits, then we require about $\sqrt{16} = 4$ computational steps. And if we double the size of the key to 8 bits, then we will require $\sqrt{256} = 16$ computational steps.

Even if we use the best known classical algorithm to find the prime factors, the *general number field sieve*, it still takes an exponentially large number of computational steps:

$$O(e^{(64/9)^{1/3}\ Log^{1/3}(r)\ (LogLog(r))^{2/3}})\ . \tag{6.43}$$

Therefore, by increasing the size of the secret key, we can easily obtain a key which will require an unfeasible amount of computational resources. This means that RSA is reasonably secure. For example, in RSA129 the problem involves the factorization of an integer with 129 digits and was the industry standard in the early 1990s. If we have a machine that runs at 1 million instructions per second, it would take about 5,000 years to break RSA129. If we have one of the latest Intel processors, then we could find the prime factors in 5 years. Therefore, a large parallel computer could break it in a matter of minutes. However, if we increase the size of the key to 200, then we would require over 2.9 billion years to find the prime factors. Therefore, we could safely presume that today nobody could have enough computational resources to break RSA200.

This way, we can make an asymmetric cipher reasonably secure. Although in this case, it is not perfectly secure, but only *computationally secure* (it makes assumptions about the computational resources available to the adversary). However, once again we have to make clear that, to date, there is no theoretical lower bound for the factorization problem. That is, at anytime in the future someone could come up with an efficient classical algorithm that factorizes large integers in polynomial time.

At the same time, within the context of the factorization problem, if we are provided with both prime factors, then to computer their product is rather trivial. Thus, this type of problem, encoded as a mathematical function, is said to be a *one-way function*. That is, it is easy to compute it one way (given both factors we can easily compute the product), but not the other (given the product it is difficult to calculate the prime factorization).

The factorization problem is also said to have a *trap door*. This means that even though the calculation of both prime factors is difficult, if somehow we are given one of them, then the problem is trivial. Indeed, if we have r, and also p, then it is very easy to compute q. In this case, knowledge of one of the factors is the trap door. This is important because the trap door is only known to the receiver of the secret message, and he uses it to decrypt his message in a short period of time. Of course, in real life applications the trap door is much more sophisticated than this, involving modular arithmetic of p and q [2].

Because the Hidden Subgroup Problem is very difficult to solve with classical resources, its instances are often used as one-way functions with trap doors for cryptographic applications. For example, RSA is a cipher that uses the periodicity instance, which is related to the factorization problem, to encode secret information. Other widely used ciphers include the El-Gamal and Diffie Hellman, which are encryption protocols based on the discrete logarithm instance of the Hidden Subgroup Problem [46]. As there is no proof that there is no classical algorithm able to solve the Hidden Subgroup Problem in polynomial time, the security of these ciphers is based on an unproved mathematical assumption. Therefore, the security of these asymmetric cryptosystems is classified as computationally secure.

However, as we discussed on the previous section, any instance of the Hidden Subgroup Problem can be solved using Quantum Fourier Transforms on a quantum computer. Furthermore, the quantum solution is exponentially faster than the best known classical method. This means that in the presence of quantum computers, all the ciphers based on instances of the Hidden Subgroup Problem will be liable to be broken in a short amount of time by someone conducting a sophisticated quantum attack. That is, all of our current standards of security are threatened by quantum technology.

The case of RSA, for example, can be broken using the famous Shor's algorithm [44, 45]. Perhaps the most important result so far in the area of quantum computing, Shor's algorithm provides a method to break RSA exponentially faster than with the best known classical method. Here, the problem of prime factorization, which is the backbone of RSA, is reduced to a periodicity problem

[2] A good description of the RSA cipher can be found in [46].

using some basic results from number theory (Euclid's Theorem, Euler's Theorem, and Euclid's Algorithm). For a large coprime integer N, this process takes $O(Log(N))$ computational steps. Then, the algorithm uses a Quantum Fourier Transform in $O(Log^3(N))$ steps to find the period of the function, as described on a previous section. Therefore, the overall complexity of Shor's algorithm is $O(Log^3(N))$, which is exponentially faster than the general number field sieve running with complexity $O(poly(N))$.

At this point is very natural to try to ask how much time will take to break RSA using a quantum computer. Unfortunately, we cannot answer this question because we do not know the running time coefficients that will characterize the first generations of quantum hardware. That is, we know that quantum computing offers a superior algorithmic complexity than classical computing, but we cannot predict at this time the running times. We can estimate, however, that if the running time coefficients of quantum hardware are of at least some MHz, then, instead of taking billions of years to break RSA200, this could be accomplished in a matter of days or weeks.

Nevertheless, we can work out a basic estimate of the scale of times involved [2]. Let us suppose we have access to a theoretical 1 Peta-Hertz classical processor. If we presume that computer technology will continue to evolve following Moore's Law, we can expect this processor sometime after the year 2044. This futuristic processor still would requires over 300 million years to break an RSA key made of 2048 bits. On the other hand, let us suppose we have access to a comparatively slow 100 MHz quantum computer with $\approx 10,000$ qubits and 2×10^{11} quantum gates. If we run Shor's algorithm using this theoretical quantum computer, we could break this cipher in less than an hour (of course, it is still difficult to predict when quantum technology will enable this size of hardware).

Before concluding this chapter, it is worth considering that, even though quantum computers can potentially break RSA, Diffie-Hellman, and El-Gamal ciphers, there may be the case of asymmetric ciphers resilient to a quantum attack. Indeed, to date, we only know how a quantum computer can be used to break those asymmetric ciphers that are based on some instance of the Hidden Subgroup Problem. In addition, there is no information, whatsoever, as to how to use a quantum computer to break a cipher that uses other types of one-way functions. So, in principle someone could develop secure, robust, and efficient asymmetric ciphers, which can withstand a sophisticated quantum attack.

Such a possibility has been explored in RSA labs [48]. Most recently RSA has proposed a rather sophisticated cryptosystem based on the Lamport digital signature using Merkle trees, which does not involve an instance of the Hidden Subgroup Problem, and, therefore, appears to be immune to a quantum attack. Unfortunately, the actual implementation of this cipher is very complex and inefficient, which makes it unfeasible. But then again, in the future such quantum-resistant ciphers could be designed to be efficient and easy to implement on existing classical hardware.

6.5 SUMMARY

Even though the Quantum Fourier Transform offers an exponential speedup over the Fast Fourier Transform, there are many limitations that impede their use for the quantum acceleration of problems in areas such as optics, acoustics, and signal analysis. Therefore, we need to look for applications where there is no need to export the entire Fourier transformation of a set of points. Such is the case of the problems that involve instances of the Hidden Subgroup Problem.

A very subtle application of Quantum Fourier Transforms, the general technique discussed in this chapter can be used to solve the periodicity, phase estimation, and discrete logarithm problems exponentially faster than with the best known classical methods. The relevance of this result is that most cryptosystems currently used today rely on some instance of the Hidden Subgroup Problem to guarantee their computational security. Therefore, the existence of quantum hardware could severely compromise the secret communications that are used day to day in the military, financial, and personal spheres. By large, this is the most important application of the quantum computing model.

Furthermore, as we will see in the next chapter, the hardware implementation of a Quantum Fourier Transform requires a small circuit. That is, it is feasible to have a quantum circuit that performs Shor's algorithm.

CHAPTER 7

Circuit Complexity Analysis of Quantum Algorithms

This chapter is concerned with the rigorous circuit complexity analysis of quantum algorithms. We will show that many complexity analyses of quantum algorithms do not consider the exponentially large number of quantum gates necessary for their circuit implementation in hardware. This suggests the possibility that the seeming computational parallelism offered by a large number of quantum algorithms is actually effected by gate-level parallelism in the reversible implementation of the quantum operator. In other words, when the total number of logic operations is analyzed, these quantum algorithms may not be more powerful than their classical counterparts. This fact has significant implications on the type of quantum algorithms that can be implemented efficiently in hardware.

7.1 QUANTUM PARALLELISM

As we discuss in Chapter 2, the power of the quantum computing model derives from a presumption that it is possible to store and simultaneously manipulate an exponentially-large amount of information in a quantum register (i.e., the amount of information is exponential in the size of the register) to perform useful computations. This would be achieved using a register of n-qubits in which information is stored in the form of a quantum superposition of $N = 2^n$ states.

More specifically, a single unitary operation can be used to transform an entire superposition of states in one logical step, which seems to imply that $O(2^n)$ state transformations occur in parallel. However, the computational use of this exponentially large level of parallelism is obstructed by the constraints of quantum theory, e.g., the collapse of the superposition induced by a measurement and the No-Cloning Theorem. Even so, as we have seen, certain quantum algorithms relying on amplitude amplification or quantum Fourier Transform techniques have been proved to exploit this parallelism to achieve optimal time performance.

In any event, even if a quantum algorithm requires fewer iterations than the best classical alternative, it is necessary to examine the relative computational expenditures per iteration because of the complexity costs associated with the reversible implementation of quantum operators [34].

7.2 ALGORITHMIC EQUITY ASSUMPTIONS

The vast majority of algorithmic analyses in the literature rely on multiple simplifying assumptions, which are not explicitly stated. In most cases, such assumptions are inconsequential because a more rigorous and complete analysis will yield essentially the same result. However, it is well known that it is possible to exploit common assumptions to achieve spurious complexity results that are not consistent with those of a more complete analysis.

For example, Shamir showed that an assumption that arithmetic operations on integers have $O(1)$ complexity can be exploited to seemingly offer an $O(\log N)$ algorithm for integer factorization - which, of course, would be more efficient even than Shor's quantum algorithm [42]. Shamir achieves this result by essentially packing a superlinear (in N) number of bits of information into a single integer while treating operations on that huge integer as still having $O(1)$ complexity. Clearly, the true complexity must take into account the scaling of the number of logic gates required to support unit-cost operations for the size of the integers necessary to perform the computations. If the number of gates grows with N (i.e., the number to be factored) then it becomes clear that the algorithm is benefiting from a degree of gate-level parallelism that is not actually available on a conventional computer. Alternatively, if it is explicitly assumed that the required gate-level parallelism actually is available, then competing algorithms may be able to exploit the same assumption to achieve comparable complexity reductions. In other words, equity of assumptions is necessary when comparing the computational complexities of different algorithms.

7.3 CLASSICAL AND QUANTUM CIRCUIT COMPLEXITY ANALYSIS

Within the context of classical computing, it is well known that the three binary operations AND, OR, and NOT, form a universal set of computational gates. Furthermore, Claude Shannon revealed in his celebrated 1949 paper that the computation of arbitrary Boolean functions requires circuits of exponential size [43]. More specifically, Shannon showed that the computer simulation of such an arbitrary Boolean function using an universal set of gates requires a classical circuit of size $O(2^n/n) = O(N/\log N)$.

However, it is important to note that most Boolean functions found in practice, like those used for arithmetic operations, require only linear-size circuits, $O(n) = O(\log N)$. As a matter of fact, it appears that explicit examples of classical Boolean functions requiring superlinear circuits are difficult to construct, and, in fact, no explicit example has been found [27, 28, 29]. Furthermore, it is observed (and conjectured as a general result) that all problems that can be decided in $O(2^n)$ time may be decided by a family of linear-size classical circuits [27].

On the other hand, in the quantum case, the model assumes that operations can be applied in parallel to all N states in a quantum superposition to solve a given algorithmic problem more efficiently than is possible, using classical hardware. However, quantum mechanics imposes a significant restriction: the transformations applied to the quantum register must be unitary. This is necessary because a nonunitary operator is equivalent to performing a measurement and thus will cause a collapse of the superposition. Fortunately, this fact does not limit the generality of quantum computation because in theory an arbitrary Boolean function can be implemented using unitary operators.

Therefore, in strong contrast to the classical case, a universal basis for quantum circuits implementing arbitrary unitary transformations has uncountably many gates. In particular, Di Vincenzo showed in 1994 that the set of all quantum gates operating on 2 bits forms a universal set of gates [53]. Further analysis determined that the number of 2-bit gates necessary to implement an arbitrary quantum operation is $O(n^3 4^n)$ [3]. A tighter bound was determined to be $O(n4^n)$ [30], which was subsequently improved to $O(4^n)$ [26]. These improvements do not affect the exponential complexity, and $O(4^n)$ appears to be the optimal bound.

7.4 COMPARING CLASSICAL AND QUANTUM ALGORITHMS

Therefore, the implementation of an arbitrary unitary operator requires $O(4^n) = O(N^2)$ 2-qubit elementary quantum gates, and the implementation of an arbitrary Boolean function requires $O(n)$ classical gates. As a consequence, if we compare the complexities for the worst case scenarios, we observe that the classical circuit requires a factor of $O(N \log N)$ fewer gates than the quantum circuit. This extra factor of $O(N \log N)$ gates required for the unitary form changes the relative resource allocation, so an alternative classical algorithm should assume the availability of the same number of logic gates. However, this is precisely what is needed to achieve the same level of parallelism as is potentially available from N states in a superposition.

Thus, the previous argument makes evident that in the worst case scenario, as showcased by the asymptotic upper bounds, classical computing requires smaller circuits than quantum computing. Furthermore, in such worst case scenario, if the classical algorithm has access to an exponentially large circuit in a parallel architecture, then it will achieve a level of parallelism equivalent to the quantum model, and without the pragmatic nuisances imposed by the destructive nature of quantum measurement and the No-Cloning Theorem.

Instead of the worst case scenario, let us now analyze a typical problem, such as an exact match query search using Grover's algorithm and classical brute force. In this case, smaller quantum circuit complexities have been found for the implementation of the oracle in Grover' algorithm. For example, it has been shown that the reversible circuit implementation of an n-qubit quantum oracle comparing a key of length n requires $\Omega(2^n)$ quantum gates [51]. Because key comparison is the simplest operation that distinguishes a unique solution, this result implies that Grover's algorithm must use an exponential (in n) number of gates to achieve its claimed time complexity.

What is important to note is that a classical algorithm can perform a key comparison in $O(n)$ time simply by comparing each of the n-bits. This means that if we were to apply $2^n/n$ classical comparisons in parallel to the 2^n states in the dataset using $O(2^n)$ gates, the solution would be found in $O(n)$ time, which is exponentially faster than Grover's $O(2^{n/2})$ complexity. In terms of the number of states $N = 2^n$, if the same number of logic gates is used by Grover's algorithm and by the classical solution, Grover's algorithm has $O(N^{1/2})$ sequential time complexity and the classical solution has $O(\log N)$ complexity. The critical observation that can be made is that this general algorithmic technique can be applied for any arbitrary oracle, and it can be concluded based on the results discussed above that there can be no case in which Grover's algorithm yields a better complexity.

Indeed, the exact same analysis and conclusions hold for the case in which the quantum register stores a superposition of indices which reference keys in an external dataset. In fact, just the $O(2^n)$ parallel gates required by Grover's algorithm to reference the external array are sufficient for the classical algorithm to achieve its complexity bound.

Thus, if we base the performance analysis of Grover's algorithm on an equity of assumptions context, we find that the classical brute force solution provides the solution to the query problem in optimal time. It is important to remark, however, that the optimality of Grover's algorithm has been proved with respect to the number of queries to the oracle, in this case, $O(N^{1/2})$. The classical solution is clearly not optimal on the number of queries of the oracle, as it requires $O(N)$, but these queries can be performed in parallel if the hardware is available.

A question that remains is whether Grover's algorithm (or any other quantum algorithm) can achieve a complexity advantage from a time/space complexity tradeoff. More specifically, changing the available number of gates affects the time complexity for both the quantum and classical algorithms, so it may be possible that there is a number of gates for which the complexity of the quantum algorithm is superior to that of the classical algorithm. This would require an exponential reduction in gates with only a sublinear increase in the time complexity of Grover's algorithm, but the analyses of time/space tradeoffs for reversible computation by Vitanyi [54] show that this is not possible.

The previous analysis can be generalized beyond Grover's algorithm to nearly any other quantum algorithm that requires an oracle. That is, a wide variety of quantum algorithms that have been proved optimal on the number of queries to an oracle, they require an exponentially large number of quantum gates. Under the equity of assumptions framework, a competing classical algorithm should be allowed access to a classical circuit of exponential size. Under these circumstances, it remains to be decided which quantum algorithms are really superior to the competing classical algorithms.

It is important to note that such exponentially large circuit complexity, as well as, the infinitely large universal set of quantum gates that are needed for the exact representation of an arbitrary unitary operator, have motivated the consideration of approximate decompositions of arbitrary unitary operators using a finite set of elementary gates. In this case, invoking the Solovay-Kitaev Theorem, it can be shown that an m-gate quantum circuit made of CNOT gates and 1-qubit gates can be realized with a circuit of size $O(m \log^c(m/\delta))$ with precision δ exclusively using only Hadamard, phase, CNOT, and $\pi/8$ gates [40]. Also, using a similar finite set of quantum gates, an arbitrary unitary operation in n-qubits can be accomplished with $\Omega(2^n \log(1/\delta)/\log(n))$ elementary gates [40]. While these approaches are successful in the reduction of the number of gates in the universal set, which is extremely important for the physical realization of quantum computers and architectures, they still imply exponentially large circuits and exponentially large pre-processing times.

Furthermore, the error in the approximation of unitary gates accumulates linearly. This means that errors will dominate the result of a quantum computation that runs for $\Omega(2^n)$ iterations. In addition, because the approximation error is reflected on the design of the target unitary operator that needs to be implemented in hardware, these errors are not correctable by any type of quantum error correction protocol. While this difficulty does not apply to all quantum algorithms (e.g., Grover's or Shor's), it certainly precludes quantum solutions or quantum implementations for any hard classical problems.

It must be emphasized that establishing that quantum computing requires $O(4^n)$ gates in the worst case does not resolve the question of whether or not quantum computation offers new computational power. It could be the case that a particular quantum algorithm is able to exploit the special structure of its associated unitary operator so that the implementation of the operator requires only a small number of gates. Shor's quantum factorization algorithm is an example in which the unitary operator can be implemented efficiently and leads to an overall complexity which exceeds that of the best openly-published classical alternative [8].

7.5 SUMMARY

It is important to remark that the observations presented in this chapter do not challenge the theoretical demonstrations of the correctness of quantum algorithms. But our research questions if they are more computationally efficient than classical alternatives under a rigorous complexity analysis that takes into account the size of the quantum circuits necessary for their implementation. That is, even if a quantum algorithm is proved to be correct, its asymptotic circuit complexity is what ultimately will decide if its hardware implementation is feasible in a scalable quantum computer.

CHAPTER 8

Conclusions

In this manuscript we have described the features of quantum theory that potentially support a more powerful model of computation than that of the classical Turing model. Specifically, we have described how a unitary operator can be applied to quantum superposition of states to transform the entire set of states in parallel. This so-called quantum parallelism is what potentially permits quantum algorithms to achieve computational complexities superior than those of any classical alternatives. Grover's quantum search algorithm is an example of a fundamental quantum technique that has applications in a wide range of compute-intensive applications ranging from databases to computational geometry to virtual reality. Shor's quantum factoring algorithm is much more limited in its range of applications but no less important in its theoretical and practical significance.

Although quantum theory seems to unambiguously imply the reality of quantum parallelism, there remains a question of whether quantum parallelism implies computational parallelism distinct from classical parallelism. Specifically, when an operator is applied to a superposition of states, the *effect* is equivalent to the application of the operator to each of the states in parallel, and this is all that is necessary to establish the correctness of quantum algorithms, such as, Grover's. As discussed in Chapter 7, however, the requirement that the operator be unitary has implications that cannot be ignored. In particular, it is entirely possible that, for certain quantum algorithms, the apparent computational savings obtained from the application of the operator to states in a superposition are always precisely balanced by the increased number of reversible gates required to implement the operator. In conclusion, although there may be some unresolved issues regarding quantum parallelism, there is no question that the future of quantum computing is the future of computing itself.

Bibliography

[1] R. Jozsa. Quantum factoring, dicrete logarithms and the hidden subgroup problem, *IEEE Computing in Science and Engineering*, 2008.

[2] C.S. Calude and G. Paunn. *Computing with Cells and Atoms: An Introduction to Quantum, DNA, and Membrane Computing*, Taylor and Francis, 2001.

[3] A. Barenco et al. Elementary gates for quantum computation. *Phys. Rev. A, 52, 3457*, 1995. DOI: 10.1103/PhysRevA.52.3457

[4] A. Pardomo et al. On the construction of model hamiltonians for adiabatic quantum computing and its application to finding low energy conformations of lattice protein models. *Physical Review A, 78, 012320*, 2008. DOI: 10.1103/PhysRevA.78.012320

[5] M.N. Vyalyi, A. Yu. Kitaev, and A.H. Shen. *Classical and Quantum Computation*. American Mathematical Society, 1999.

[6] A. Andersson and K. Swanson. On the difficulty of range searching. *Computational Geometry with Applications*, 8(3):115–122, 1997. DOI: 10.1016/S0925-7721(97)00005-9

[7] G. Van Assche. *Quantum Cryptography and Secret Key Distribution*. Cambridge University Press, 2006.

[8] St. Beauregard. Circuit for shor's algorithm using 2n+3 qubits. *Quantum Information and Computation*, 3(2), 175, 2003. DOI: arXiv:quant-ph/0205095v3

[9] M. Boyer, G. Brassard, P. Hoyer, and A. Tapp. Tight bounds on quantum searching. *Proceedings of the Fourth Workshop on Physics and Computation*, 1996. DOI: 10.1002/(SICI)1521-3978(199806)46:4/5<493::AID-PROP493>3.0.CO;2-P

[10] G. Brassard, P. Hoyer, M. Mosca, and A. Tapp. Quantum amplitude amplification and estimation. *e-print quant-ph/0005055*, 2000. DOI: arXiv:quant-ph/0005055v1

[11] E. Oran Brigham. *The Fast Fourier Transform*. Prentice-Hall, 1974.

[12] M.F. Cohen and J.R. Wallace. *Radiosity and Realistic Image Synthesis*. Morgan Kaufman, 1993.

[13] ed. D. Bouwmeester. *The Physics of Quantum Computation*. Springer, 2000.

[14] P.A.M. Dirac. *The Principles of Quantum Mechanics, Fourth Edition*. Oxford University Press, 1982.

[15] C. Durr and P. Hoyer. A quantum algorithm for finding the minimum. *e-print quant-ph/9607014*, 1999.

[16] M.C. Loui, E. Allender, and K.W. Regan. Complexity Classes, in *Algorithms and Theory of Computation Handbook*, (Hardcover). CRC-Press; 1st edition (September 30, 1998).

[17] R.P. Feynman. Simulating physics with computers. *Int. J. Theo. Phys. 21:467*, 1982. DOI: 10.1007/BF02650179

[18] J.D. Foley, A. Van Dam, S.K. Feiner, and J.F. Hughes. *Computer Graphics, Principles and Practice*. Addison-Wesley, 1996.

[19] A. Glassner. *An Introduction to Ray Tracing*. Academic Press, 1989.

[20] A. Glassner. Andrew Glassner's notebook: Quantum computing, part 1. *IEEE Computer Graphics with Applications*, 21(4), Jul/Aug 2001. DOI: 10.1109/38.953464

[21] A. Glassner. Andrew Glassner's notebook: Quantum computing, part 2. *IEEE Computer Graphics with Applications*, 21(5), Sep/Oct 2001. DOI: 10.1109/38.946635

[22] A. Glassner. Andrew Glassner's notebook: Quantum computing, part 3. *IEEE Computer Graphics with Applications*, 21(6), Nov/Dec 2001. DOI: 10.1109/38.969611

[23] L. Grover. A Fast Quantum Mechanical Algorithm for Database Search, *Proc. of 28th ACM Annual STOC*, 212–219, 1996. DOI: 10.1145/237814.237866

[24] L. Grover. Quantum computers can search rapidly by using almost any transformation. *lanl e-print quant-ph/9712011*, 1997. DOI: 10.1103/PhysRevLett.80.4329

[25] H. Wang et al. Quantum algorithm for obtaining the spectrum of molecular systems. *Physical Chemistry, Chemical Physics. In press*, 2008. DOI: 10.1039/b804804e

[26] M. Mottonen, J.J. Vartiainen, and M.M. Salomaa. Efficient decomposition of quantum gates. *Phys. Rev. A, 92, 177902*, 2004. DOI: 10.1103/PhysRevLett.92.177902

[27] V. Kabanets and JY Cai. Circuit minimization problem. *Proceedings of the Thirty-Second Annual ACM Symposium on Theory of Computing*, 2000. DOI: 10.1145/335305.335314

[28] R.J. Lipton. Some consequences of our failure to prove non-linear lower bounds on explicit functions, *Proceedings of the Structure in Complexity Theory Conference*, 1994. DOI: arXiv:quant-ph/9508006v1

[29] S. Aaronson. Oracles are subtle but not malicious, *Proceedings of IEEE Complexity*, 2006. DOI: doi:10.1117/12.541624

[30] E. Knill. Approximation by quantum circuits. *LANL Report LAUR-95-2225*, 1995. DOI: 10.1117/12.602928

[31] M. Lanzagorta and J. Uhlmann. Quantum computational geometry. In *Proceedings of the Quantum Information and Quantum Computation Conference*. SPIE Defense and Security Symposium, 2004. DOI: 10.1145/1198555.1198722

[32] M. Lanzagorta and J. Uhlmann. Hybrid quantum computing. In *Proceedings of the Quantum Information and Quantum Computation Conference*. SPIE Defense and Security Symposium, 2005. DOI: 10.1117/12.778019

[33] M. Lanzagorta and J. Uhlmann. Quantum rendering. *Tutorial Presented at the Siggraph Conference*, 2005.

[34] M. Lanzagorta and J. Uhlmann. Is quantum parallelism real? *Proceedings of the Quantum Information and Quantum Computation Conference of the SPIE Defense and Security Symposium*, 2008. DOI: 10.1117/12.486410

[35] M. Lanzagorta and J. Uhlmann. Multi-object quantum search. *Submitted to Quantum Information Letters*, 2008.

[36] M. Lanzagorta, J. Uhlmann, and R. Gomez. Quantum rendering. In *Proceedings of the Quantum Information and Quantum Computation Conference*. SPIE Defense and Security Symposium, 2003.

[37] D. Luebke, M. Reddy, J.D. Cohen, A. Varshney, B. Watson, and R. Huebner. *Level of Detail for 3D Graphics*. Morgan Kaufman, 2002. DOI: 10.1007/s00493-004-0009-8

[38] M. de Berg et al. *Computational Geometry: Algorithms and Applications, third edition*. Springer, 2008.

[39] M. Grigni et al. Quantum mechanical algorithms for the nonabelian hidden subgroup problem. *Combinatorica*, Vol. 24:1, 2004.

[40] M.A. Nielsen and I.L. Chuang. *Quantum Computation and Quantum Information*. Cambridge University Press, 2000.

[41] F.P. Preparata and M.I. Shamos. *Computational Geometry*. Springer-Verlag, 1985.

[42] A. Shamir. Factoring numbers in O(log(n)) arithmetic steps. *Information Processing Letters*, Vol. 8, Number 1, 1979. DOI: 10.1109/SFCS.1994.365700

[43] C. Shannon. The synthesis of two-terminal switching circuits. Bell Systems Technical Journal, 28(1):59-98, 1949 DOI: 10.1137/S0097539795293172

[44] P. Shor. Algorithms for quantum computation: Discrete logarithms and factoring. *Proc. of 35th Symp. on FOCS*, 1994.

[45] P. Shor. Polynomial-time algorithms for prime factorization and discrete logarithms on a quantum computer. *SIAM Journal on Quantum Computing*, 26(5):1484–1509, 1997. DOI: 10.1002/cphc.200400582

[46] D.R. Stinson. *Cryptography: Theory and Practice*. Chapman and Hall, CRC, 2nd Edition, 2002. DOI: 10.1007/b97182

[47] J. Stolze and D. Suter. *Quantum Computing: A Short Course from Theory to Experiment*. Wiley-VCH, 2000.

[48] M. Szydlo. Merkle tree traversal in log space and time. *Eurocrypt 2004*, 2004. DOI: 10.1103/PhysRevLett.100.160501

[49] T. Jiang, et al. Computability, in *Algorithms and Theory of Computation Handbook*, (Hardcover). CRC-Press; 1st edition (September 30, 1998).

[50] L. Maccone, V. Giovannetti, and S. Lloyd. Quantum random access memory. *Phys. Rev. Lett. 100, 160501*, 2008. DOI: 10.2200/S00144ED1V01Y200808QMC001

[51] I.L. Markov, V. Shende, A. Prasad, and J.P. Hayes. Synthesis of reversible logic circuits. *IEEE Transactions on CAD*, 2003. DOI: 10.1103/PhysRevA.51.1015

[52] S. Venegas. *Quantum Walks for Computer Scientists*. Morgan and Claypool, 2008. DOI: 10.1145/1062261.1062335

[53] D. Di Vicenzo. Two-bit gates are universal for quantum computation. *Phys. Rev. A*, 51:1015–1022, 1995.

[54] P.M.B. Vitanyi. Time, space, and energy in reversible computing. *Conf. Computing Frontiers*, 435–444, 2005.

[55] Rajagopal Nagarajan, Nikolaos Papanikolaou, and David Williams. Simulating and Compiling Code for the Sequential Quantum Random Access Machine. In P. Selinger, Proceedings of the 3rd International Workshop on Quantum Programming Languages (QPL 2005), June 30-July 1, 2005, DePaul University, Chicago, USA. Electronic Notes in Theoretical Computer Science 170, pp. 101-124.

[56] M. Lanzagorta and S. Bique, Introduction to Reconfigurable Supercomputing, course notes of the turorial presented at the Department of Defense High Performance Computing Modernization Program Users Group Conference, Seattle, WA, 2008.

[57] C. Zalka. Grover's quantum searching algorithm is optimal, *Phys. Rev.*, A60 2746-2751, 1999.

[58] M. Lanzagorta. Defense Applications of Quantum Computation, brief presented to the US Air Force Research Lab, Rome, NY, 2008.

[59] Michelangelo Grigni, J. Schulman, Monica Vazirani, and Umesh Vazirani. Quantum Mechanical Algorithms for the Nonabelian Hidden Subgroup Problem, *Combinatorica*, 24(1), 0209-9683, 137–154, 2004. DOI: 10.1007/s00493-004-0009-8

[60] V. Buzek and M. Hillery. Quantum Copying: Beyond the No-Cloning Theorem, *Physical Review A*, 54(1844), 1996.

[61] R. Werner. Optimal Cloning of Pure States, *Physical Review A*, 58(1827), 1998.

[62] V. Buzek and M. Hillery. Universal Optimal Cloning of Arbitrary Quantum States: From qubits to Quantum Registers, *Physical Review Letters*, 81(22), 1998.

[63] V. Buzek and M. Hillery. Quantum Copying: A Network, e-print quant-ph/9703046, 1997.

Biographies

Marco Lanzagorta

Dr. Marco Lanzagorta is Technical Fellow and Senior Principal Scientist at the Advanced Engineering and Sciences Division of ITT Corporation, a scientific consultant for a project for the US Naval Research Laboratory in Washington, DC, and an Affiliate Associate Professor at the Center for Quantum Studies of George Mason University in Fairfax, VA. In addition, Dr. Lanzagorta is co-editor of the Synthesis Lectures on Quantum Computing published by Morgan and Claypool. Dr. Lanzagorta received a PhD in Theoretical Physics from the University of Oxford, UK, and in the past he worked at the European Organization for Nuclear Research (CERN) in Switzerland and the International Centre for Theoretical Physics (ICTP) in Italy.

Jeffrey Uhlmann

Dr. Uhlmann is a leading researcher in the theory of stochastic tracking and control. He is currently a professor at the University of Missouri-Columbia after working for 12 years at the Naval Research Laboratory in Washington, DC. He has a bachelors degree in philosophy, a masters in computer science, and a doctorate in robotics from the University of Oxford in England. Dr. Uhlmann is also a scholar of cult and B-movies, especially of the lucha and kaiju genres.